ESWT and Ultrasound Imaging of the Musculoskeletal System

C. E. Bachmann G. Gruber W. Konermann
A. Arnold G. M. Gruber F. Ueberle

ESWT and Ultrasound Imaging of the Musculoskeletal System

Co-author: L. Gerdesmeyer

2nd revised and translated edition,
with 116 Figures in 177 separate Illustrations and 17 Tables

Responsible for the translation: H. Hermeking

STEINKOPFF
DARMSTADT

ISBN 978-3-7985-1252-8   ISBN 978-3-642-48805-4 (eBook)
DOI: 10.1007/ 978-3-642-48805-4

Die Deutsche Bibliothek – CIP-Einheitsaufnahme
A catelogue record for this publication is available from Die Deutsche Bibliothek

Steinkopff Verlag Darmstadt,
a member of BertelsmannSpringer Science+Business Media GmbH

© Steinkopff Verlag Darmstadt 2001

Production: Klemens Schwind
Cover Design: Erich Kirchner, Heidelberg
Typesetting: K+V Fotosatz GmbH, Beerfelden

SPIN 10780301   105/7231-5 4 3 2 1 0 – Printed on acid-free paper

What we know is a drop of water,
What we don't know is an ocean

Sir Isaac Newton (1643–1727)

Preface

Extracorporeal Shock Wave Therapy (ESWT) is a new method for the treatment of numerous chronic disorders of the musculoskeletal system. Most of the experience has been gained in ESWT applications for calcific tendinitis of the shoulder joint, epicondylitis of the elbow and plantar fasciitis. Further applications are other forms of chronic tendinitis, as for example jumper's knee, and the treatment of pseudarthrosis. Without completed multi-centered and well-controlled studies, and therefore without final validation, the clinical efficiency of this new therapeutic method is sometimes viewed critically. This situation is quite common, whenever a new technology or new applications are introduced in medicine. Despite this, ESWT has spread widely in the global orthopedic community. ESWT has proven to be particularly beneficial in the treatment of chronic pain as an alternative treatment prior to surgical intervention.

This book has three objectives:
1. To describe presently performed orthopedic applications of ESWT.
2. To demonstrate the application of ESWT using ultrasound imaging.
3. To present standard sectional planes for ultrasound examination of joints (shoulder, elbow, knee and ankle joints including the foot) defined by the work group Musculoskeletal System of the German Society for Ultrasound in Medicine (DEGUM). Application of ultrasound transducers, sonographic anatomy and additional indications for ultrasound imaging are included.

By linking ESWT and sonography we hope to facilitate the approach to this topic and the handling of the device for those colleagues, who have not previously used this new method. For more experienced physicians the book is intended to serve as a reference, an application manual and an ultrasound refresher. Obviously this book can offer basic guidelines only, since the procedures have to be acquired by continuously practicing them-in the beginning under supervision. For the latter, ESWT application courses and sonography courses are recommended.

The present book has been revised and translated from its first edition, which was published in German. In addition, a review of most recent study results has been included, and the referenced literature has been updated.

We thank Dornier Medizintechnik GmbH, Wessling for the excellent support during the development of this book.

Special thanks go to Ms. Dr. med. G. Volkert and Ms. B. Riegel, Steinkopff Verlag, for the superb, always friendly and understanding cooperation.

We thank the Steinkopff Verlag for the high quality preparation of this book.

We thank Ms. G. Paßlack, medical photographer of the Orthopädische Klinik der Justus-Liebig-Universität Gießen for her photographic support.

September 1998,   Christine E. Bachmann, Germering
revised January 2001   Gerd Gruber, Heidelberg
                        Werner Konermann, Hessisch Lichtenau
                        Astrid Arnold, Wessling
                        Gabi M. Gruber, Heidelberg
                        Friedrich Ueberle, Wessling

Table of Contents

List of Authors

Dr. med.
CHRISTINE E. BACHMANN
Orthopedic Practice
Kurfuerstenstrasse 2
D-82110 Germering, Germany

Priv.-Doz. Dr. med.
GERD GRUBER
ATOS Clinic
Bismarckstrasse 9 – 15
D-69115 Heidelberg, Germany

Priv.-Doz. Dr. med.
WERNER KONERMANN
Lichtenau e.V.
Clinic for Orthopedics
and Traumatology
Am Muehlenberg
D-37235 Hessisch Lichtenau
Germany

ASTRID ARNOLD
Sector Foresta, 23-7 B
E-28760 Tres Cantos (Madrid)
Spain

Dr. med.
GABI M. GRUBER
ATOS Clinic
Bismarckstrasse 9 – 15
D-69115 Heidelberg, Germany

Dr. Ing. FRIEDRICH UEBERLE
Dornier Medizintechnik GmbH
Argelsrieder Feld 7
D-82234 Wessling, Germany

Dr. med.
LUDGER GERDESMEYER
Clinic for Orthopedics
and Sports Orthopedics
Technical University Munich
Ismaninger Strasse 22
D-81675 Munich, Germany

Übersetzung:
Dr. HAJO HERMEKING
Dornier Medizintechnik GmbH
Argelsrieder Feld 7
D-82234 Wessling, Germany

1 Introduction

In urology, extracorporeal shock wave lithotripsy is considered to be the golden standard in stone therapy[1]. This worldwide revolutionary method has almost completely replaced invasive surgery. A similar promising development is emerging for specific orthopedic indications.

Since the early 1990s effects of shock wave application have been used in orthopedics for the treatment of osseous tissue (pseudarthrosis) and soft tissues adjacent to bones. In the treatment of pseudarthrosis the osteogenic effect is used to induce callous tissue growth. Other applications are the treatment of enthesopathies, which are painful changes in the region of osseous insertions of tendons and fascia, sometimes accompanied by calcifications, or more general the treatment of tendinitis. The treatment is centered on the transitional region between tendon (or fascia)/tendon sheath and bone/periosteum. For shock wave treatment of chronic tendinitis, such as lateral and medial epicondylitis, plantar fasciitis and calcific tendinitis positive midterm results have also been presented[2]. A review of the most recent study results has been added to chapters 8 to 11.

It is important to point out that for the latter indications pain treatment may be performed with "soft" ESWT, using low to medium energy levels, calcium deposit disintegration is observed at medium to high energy levels, while high energy shock waves are used to induce osteogenesis. The treatment with low to medium energy levels causes changes in the membrane permeability of cells in the focal zone and neurogenic hyperstimulation effects leading to analgesia as well as metabolic changes, which may also contribute to absorption of calcium deposits.

Serious complications associated with the application of ESWT can be avoided by observing recommended indications and contraindications, optimal localization and correct coupling of the therapy head.

In the future, treating physicians should be certified prior to administering ESWT.

---

[1] Chaussy et al, 1993
[2] Rompe, 1996; Loew, 1993

According to current clinical experience, shock wave therapy is a promising, minimally invasive alternative to surgery for specific orthopedic indications. Factors supporting the application of shock wave therapy are the low rate of complications, mostly out-patient treatment, frequently limited (local or systemic) anesthesia or no use of anesthesia, the cosmetically positive outcome (no scars) and the remaining option for surgical intervention at a later time.

1.1 ESWT in Orthopedics

In 1997 and 1998 about 60,000 patients were treated with orthopedic shock wave therapy annually in Germany. Since then this number has decreased due to a change in reimbursement conditions, while the method has started to be introduced internationally. To meet reimbursement requirements in Germany this new form of therapy has to be validated by multicenter, controlled, double blind studies. These studies are presently being performed in Germany and – for FDA approval purposes – in the USA.

The prerequisite for the application of ESWT always has been the patient's failure to respond to conservative treatment.

ESWT is a completely new form of therapy which has to be positioned between former unsuccessful conservative therapy and surgery.

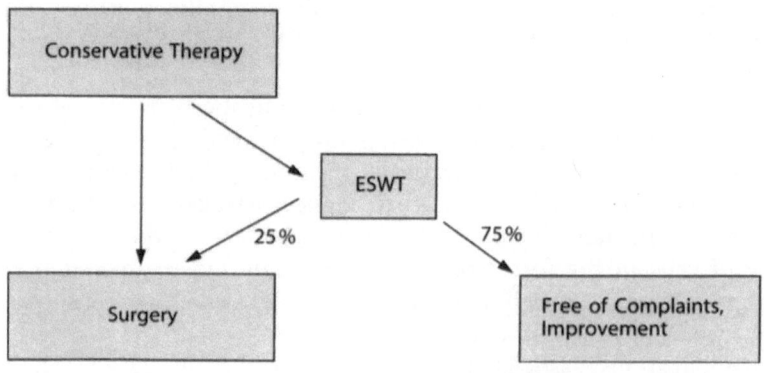

Fig. 1.1. The role of ESWT

When compared to conventional surgical treatment, shock wave therapy offers distinctive advantages for both, patients and physicians.

Advantages of ESWT
1. Out-patient treatment
2. Short treatment time
3. Reduced risk as compared to surgery
4. No general/partial anesthesia (local anesthesia as an exception)
5. No special post-treatment
6. Cosmetically optimal solution (no scars)
7. When used in sports medicine: Not classified as doping
8. Adequate patient care in residential practices
9. Extended services by practices
10. Low cost alternative for surgery
    - no hospital costs
    - no post-treatment costs
    - shortened treatment time
    - shortened period of disability for work.

2 Introduction to Physical Principles of Shock Waves

2.1 Discovery of Shock Wave Applications in Medicine

In the early 1970s engineers at Dornier GmbH, one of the leading aircraft and space industries of Germany, were studying material damage at the site of impact of rain drops on the surface of airplanes flying at supersonic speed. The damage was caused by generation of shock waves and their interaction with the surface of the airplane. The observations stimulated the idea to use this effect for medical purposes. In 1980 the first human was treated for kidney stones incorporating this effect using the first shock wave lithotriptor, the Dornier HM1 [1]. The first serial shock wave lithotriptor, the Dornier HM3, was brought to the market in 1983.

2.2 Wave Patterns

Shock waves used in lithotripsy and in ESWT are pulsed waves. This is in contrast to acoustic waveforms used in many other medical applications, which are mainly continuous waveforms.

Continuous high intensity ultrasound waves primarily generate heat in the body. This effect is used for therapeutic applications. To avoid overheating, the sound wave intensity is limited to low levels (3 W/cm^2).

For diagnostic ultrasound applications short wave pulses are used, with the number of oscillations kept as low as possible. In this case the intensity is also restricted to avoid the possibility of side effects.

Contrary to this, ESWT uses high intensity pulses with a low repetition rate (maximum 240 pulses per minute). The temperature rise in the focus region is negligible as far as therapeutic effects or side effects are concerned. The effects of shock waves important for ESWT result from the mechanical load on the tissue.

---

[1] For a more detailed description see
Ueberle, in Siebert, Buch, 1998 and
Ueberle, in Chaussy et al, 1998

Continuous Sine Wave (e.g., Hyperthermia)

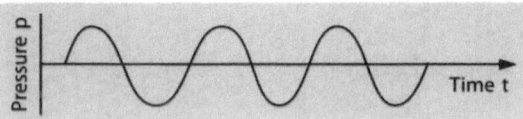

Fig. 2.1. Continuous sine wave

Sound Pulse (e.g., Diagnostic Ultrasound)

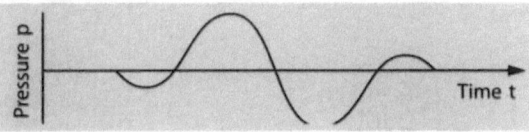

Fig. 2.2. Sound pulse

Shock Wave

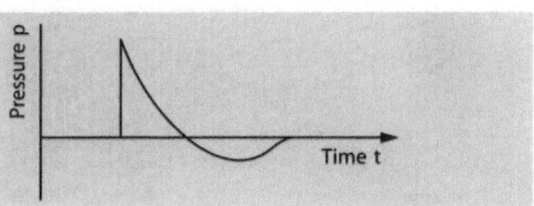

Fig. 2.3. Shock wave

2.3 Shock Wave Generation Systems

Shock wave generation systems consist of an electrical energy source, an electro-acoustic conversion mechanism and a device to focus the shock wave.

The first lithotriptors used a spark source, similar in design to the spark plug of a car. This "classic" electrohydraulic source is still used in lithotriptors today. Also in use are piezoelectric shock wave sources in which the piezoelectric effect, i.e., a change in the size of the piezo-ceramic elements, when under electric charge, is used to generate a pressure pulse.

The most commonly used generation principle today is the electromagnetic shock wave generation principle.

2.4 Technology of Shock Wave Generation

(See for Fig. 2.4 and Table 2.1 on page 6.)

2.5 Dornier Electromagnetic Shock Wave Emitter (EMSE)

The Dornier Electromagnetic Shock Wave Emitter (EMSE) employs the electromagnetic shock wave generation principle (Fig. 2.5). The EMSE features a flat coil (1). When releasing a shock wave, a pulse shaped electrical current travels through this coil, generating a strong magnetic field.

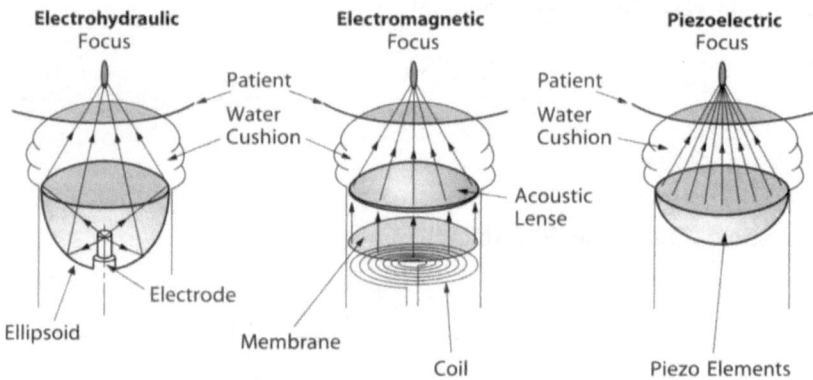

Fig. 2.4. Shock wave generation systems

Table 2.1. Technological comparison of shock wave emitting devices

Electrohydraulic Source	Electromagnetic Source	Piezoelectric Source
Generation of spherical shock waves	Generates a plane shock wave	Very large shock source
Electrode lifetime: 2–3 treatments	The entire shock wave is focused	Very sharp focusing
Discontinuous burn-off of electrode	No consumables	High energy flux density
Discontinuous energy emission	Precisely controlled energy delivery	Extremely small focal size
Varying focal position	High energy flux density	Low total energy per shock wave
Varying focal size (can vary during treatment)	Small focal size (focal size can be precisely adjusted through the geometry of the lenses)	High maintenance costs
High service costs	Constant focal size	Depending on transducer size, high negative pressure components possible
Parts of shock wave enter the body unfocused and uncontrolled	High total energy per shock wave	
	Precisely defined focus position	
	Long lifetime	

An isolated conductive membrane (2) is positioned above the coil. An eddy current is produced in this coil by the primary magnetic field, which generates a secondary magnetic field with opposite polarization. These two magnetic fields lead to repulsion of the membrane. Thereby the water in the vicinity of the membrane is compressed and a plane high pressure pulse is released, which is focused by an acoustic lens (3). The pressure across the membrane surface is essentially constant.

The EMSE generates an effective ESWT pressure pulse, which is focused by an acoustic lens. Focusing the pressure pulse leads to the development of a shock wave (Fig. 2.6) in the focal area.

The EMSE is characterized by
- Large energy range (dynamics: low, medium and high energy)
- Constant energy delivery[1]
- Precisely defined focal position[1] and focal size
- Long lifetime
- No consumables.

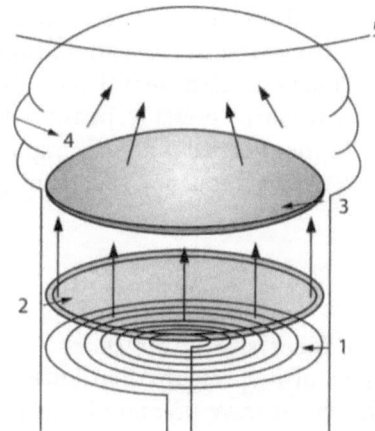

Fig. 2.5. EMSE (Electromagnetic Shock Wave Emitter):
1 Coil, 2 Membrane, 3 Acoustic lens, 4 Coupling cushion,
5 Surface of the patient's body

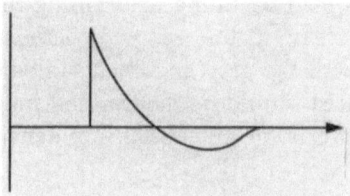

Fig. 2.6. Shock wave

2.6 Properties of Shock Waves

Shock waves[2] are mechanical waves, which travel in matter in any of its phases, either gaseous, liquid or solid. For medical purposes shock waves are generated in water and introduced into the patient's body of comparable acoustical consistency. To allow entry into the patient's body a coupling medium is mandatory (ultrasound gel, etc.).

Pressure amplitudes currently used in therapeutic applications range from 8 MPa to more than 100 MPa (1 MPa = 10 bar, equal to 10 times atmospheric pressure).

Shock waves travel with a velocity greater than the velocity of sound. As used in medical applications, shock waves only exceed the velocity of sound by a very small amount. Therefore these shock waves can be described as pressure pulses, which are very short time sound waves. At the wave front of the shock wave, the positive pressure rises extremely rapidly from ambient to maximum pressure, followed by a short phase of negative pressure.

---

[1] S. Schraebler (99)
[2] For recent literature on shock waves see Ueberle (95, 97, 99, 2001), Wess et al. (95, 97), Maier et al (98), Delius (98)

Shock waves applied in medicine are characterized by

- Rapid pressure rise (less than 10 ns)
- Extremely high maximum pressure (8 to more than 100 MPa)
- Short duration (less than < 10 μs)
- A negative pressure phase with a maximum amplitude typically 0.1 to 0.5 of the positive pressure amplitude
- A broad frequency spectrum (generally 200 kHz–20 MHz)

2.7 Shock Wave Focus

The focus size of the shock wave is defined by the axial and lateral dimensions of its pressure distribution, at which one half of the maximum value of the shock wave pressure is measured. The focus has the shape of an elliptical "cigar", with its long axis in the direction of shock wave propagation. The penetration depth of the shock wave focus into the patient's body is determined by filling or draining the water cushion of the therapy head.

Due to the way it is defined, the focus size does not necessarily correlate with the area, in which biological shock wave effects are to be expected. This area should be determined more appropriately by the threshold for biological effects of the energy flux density and associated energies (see below).

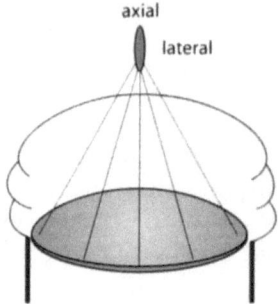

axial

lateral

Fig. 2.7. Shock wave focus

2.8 Pressure Field

To describe the pressure field the following x, y, z coordinate system is chosen. The z-axis is the symmetry axis of the therapy head running through the center of the membrane and the acoustic lens. The x- and y-axes intersect with the z-axis at the therapeutic focal point F. The shock wave field is rotationally symmetric around the z-axis, somewhat like a "cigar". For this reason only measurements along either the x- or y-axis are necessary to determine the pressure distribution of the shock wave field at a given value of z.

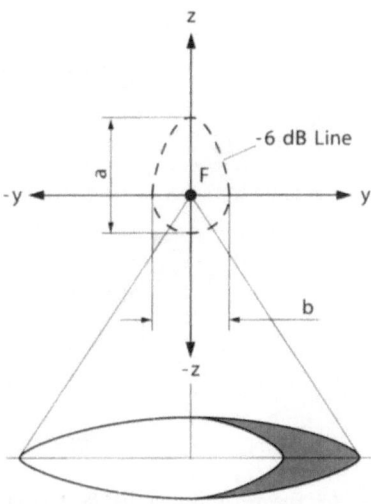

Fig. 2.8. Pressure field

2.9 Energy Flux Density and Effective Focal Energy

Energy Flux Density

Energy flux density is defined as the shock wave energy which flows through an area perpendicular to the area of propagation per unit area. The energy flux density is measured in mJ/mm^2. It is equal to the time integral over the square of the local pressure divided by the product of the density and the sound velocity. The time integral extends over the duration of the shock wave. If the time integral is limited to the duration of the positive pressure portion of the shock wave, the positive energy flux density results. The maximum value of the energy flux density is in the center of the focus. Outside the focus the lateral decay of the energy flux density is about the same as the lateral pressure decay. Biological shock wave effects are associated with threshold values of the energy flux density (See section 4.1)[1,2].

Effective Focal Energy

Effective focal energy is defined as the energy that flows through a circular plane perpendicular to the z-axis (action plane) with a diameter of 12 mm centered at the focal point (the typical size of a kidney stone). The effective focal energy results by integrating the energy flux density over the respective area. Other diameters for the action plane may be appropriate depending on the application, e.g., 5 mm (typical size of action plane chosen for pain treatment). Effective focal energies of different shock wave sources are compa-

---

[1] Rompe, 1997
[2] Steinbach, 1993

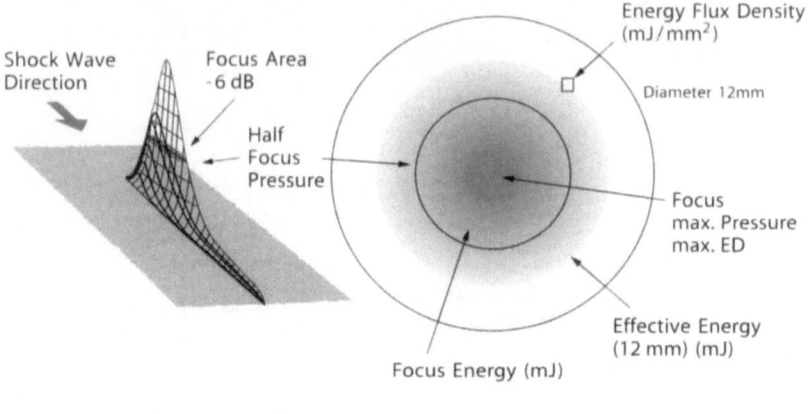

Fig. 2.9. Energy flux density and focal energy

rable only if they are associated with action planes of identical size. In the future the size of these action planes may be determined by thresholds of biological effects. The effective focal energy per pulse is measured in mJ.

2.10 Aperture Angle

The aperture angle is determined by the angle of a cone between the focal point and the opening of the shock wave source. The following general rule applies: The larger the angle of aperture of the system, the higher the focal pressure and the focal energy flux density at a given shock wave level. Lateral and axial focal dimensions decrease as the angle of aperture increases. A shock wave source with low effective energy with a large aperture angle may still have a high-energy flux density in the focal point. Therefore a comparison of different shock wave sources makes sense only if effective energies associated with action planes of identical size are compared.

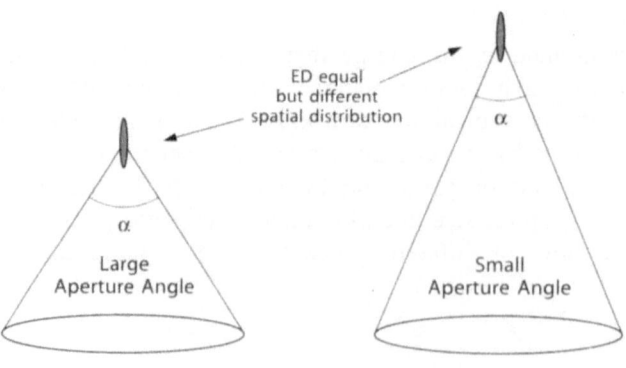

Fig. 2.10. Aperture angle

Table 2.2. Comparison of large aperture angle vs. small aperture angle

Large Aperture Angle	Small Aperture Angle
Smaller rise towards focal point	Larger rise towards focal point
Shorter distance of focus from lens	Longer distance of focal point from lens
Smaller focus	Increased focus
Higher maximum pressure	Lower maximum pressure
Higher energy flux density at the focal point (rapidly decreasing laterally, therefore possibly smaller effective focal energy)	Lower energy flux density at the focal point
	Larger focal diameter
Smaller focal diameter	

2.11 Cavitation

The high-pressure phase of the shock wave is followed by a negative pressure phase. Water is forced to undergo a local phase transition, when exposed to only a few MPa of negative pressure – water breakes up locally. Small gas bubbles are generated. This phenomenon is called cavitation. These gas bubbles are called cavitation bubbles.

The threshold for cavitation depends on the presence of cavitation seeds. Small cavitation bubbles may themselves serve as cavitation seeds for larger bubbles.

Consecutive shock waves may cause cavitation bubbles to grow to diameters of several micrometers in size. In water they can even reach millimeter size. In addition new cavitation bubbles may be generated.

The pulse duration of the negative pressure signal also plays an important role: the shorter the duration, the greater the capability of the tissue to adapt to high-tension amplitudes.

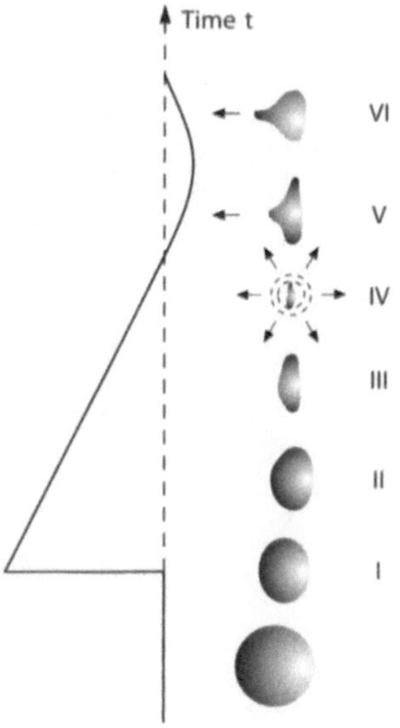

Fig. 2.11. Collapsing of cavitation bubbles.
I–III The bubble is struck and compressed by a shock wave.
IV The bubble collapses and emits a secondary shock wave of shorter duration. **V, VI** A water jet is directed toward the nearest surface through the center of the bubble

As a general rule, formation of cavitation bubbles increases with increasing negative pressure and with increasing pulse frequency of the shock waves. Cavitation bubbles may collapse either by themselves or after being struck by subsequent shock waves, generating a locally extremely high pressure pulse at the point of collapse associated with directed water jets.

Very likely cavitation bubbles are the cause of small petechial hemorrhages or hematomas under the skin at the site of the shock wave exiting from the patient's body. Cavitation bubbles also may influence the permeability of cell membranes.

2.12 Measuring Methods

Shock wave parameters are measured using standard IEC (International Electrotechnical Commission) measuring procedures. The measurements are performed by placing the probes at different positions in the focal zone of the Dornier EMSE. Today the most reliable measurements are performed with a newly developed laser fiber probe (Eisenmenger, Stuttgart, Germany).

This laser fiber probe is capable of registering both positive and negative pressure waves in the Giga Hertz range. These properties guarantee accurate measuring results of the shock wave field. In particular, the laser fiber probe reproduces the negative pressure components and therefore the effective energy levels of the total pressure field more accurately than any other known measuring methods.

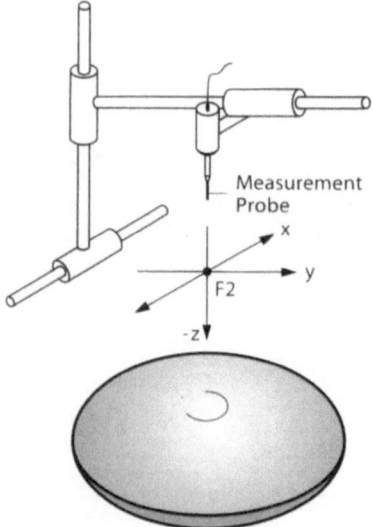

Fig. 2.12. Measuring procedures

2.13 Biological Effects of ESWT

The following biological effects are associated with the exposure of biological tissue to shock waves.

Table 2.3. Effects of ESWT in the body

Shock Wave Effect	Effect in the Body	Indication
Cavitation bubbles	Hematoma formation, nerve stimulation Effects at the cell level	Pseudarthrosis, tendinitis
	Change in the consistency of calcium deposits, fracture formation	Calcific tendinitis
Pressure/tension phases including shear forces	Analgesic effect	Tendinitis, pseudarthrosis
	Metabolic stimulation	Tendinitis, pseudarthrosis

The literature warns about the risk of possible damage by cavitation-induced free radicals.

3 Extracorporeal Shock Wave Therapy Systems

3.1 Components of a Typical ESWT System

A typical ESWT system is comprised of
- Shock wave source with controls and coupling cushion, water circulation unit
- Shock wave source positioning system
- Rack for power supply, shock wave generator and controls
- Localization device(s) or localization aids
    - Markings indicating the direction of shock wave propagation
    - Laser pointer
    - Localization device(s) using ultrasound or X-ray imaging
- Imaging systems: Ideal are isocentrically integrated systems:
    - Ultrasound system, either inline or outline
    - X-ray system.

3.2 Targeting Aids for Localization using Patient Feedback

Often the ESWT treatment site is determined with the aid of patient feedback. The physician palpates the painful region. With the aid of patient feedback, the treatment site is located and marked by an X on the skin surface by the physician.

The following localization aids are available for patient-oriented coupling:

Markings of the Shock Wave Direction on the Coupling Cushion: The markings on the coupling cushion of the therapy head indicate the direction of the z-axis relative to the patient. This serves as a visual localization aid for the physician to identify the axial position of the shock wave focal point (axial geometric focal center).

Laser Pointer: The laser pointer emits a red light beam to mark the lateral geometric center of the shock wave focus on the patient's skin. It is also a visual localization aid to determine the penetration depth of the shock wave focal point. When the tangential 2-point technique is used, ideally the laser pointer aims at the (X) marked center of pain.

Fig. 3.1. Laser pointer and marking of the shock wave direction on the coupling cushion

The markings on the coupling cushion and the laser pointer allow presentation of the focal point of the shock wave.

3.3 Localization Systems

Generally, a localization device should present the anatomy in the vicinity of the focal point in all three spatial dimensions.

3.3.1 Ultrasound

The ultrasound localization system is either called inline or outline depending on the positioning of the transducer with respect to the therapy head.

Inline Ultrasound System
In the case of an inline ultrasound localization system the transducer is located in the center of the shock source. The transducer can be turned around its axis or it can be moved in an axial direction.

It is an advantage of the inline ultrasound localization system that the ultrasound system and therapy source use the same entrance window.

It is disadvantageous that the inline transducer obstructs a portion of the shock waves, leading to artifacts in the ultrasound image. Increased primary acoustic energy has to compensate the resulting energy loss. Fixing the transducer to the therapy head also limits its flexibility. In particular, free choice of the entrance window for ultrasound sectional plane guidance is not possible. Both water and the coupling cushion impair the image quality causing additional artifacts, which in turn complicates the localization process. Since the transducer is located directly in the shock wave field, the lifetime

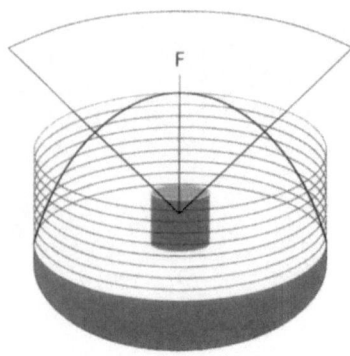

Fig. 3.2. Inline ultrasound localization system

of the transducer may be reduced significantly necessitating its early replacement.

Advantages:
1. Ultrasound image and therapy sources use the same entrance window.

Disadvantages:
1. The inline transducer reflects a portion of the therapeutic shock waves.
2. Diffraction waves are produced at the inline transducer.
3. Restricted maneuverability.
   – transducer is fixed in the therapy head.
   – entrance window not variable.
4. Water: impaired image, artifacts.
5. Lifetime shortened: Transducer is located in the shock wave field.
6. Sector transducers, generally used in inline ultrasound localization devices, lead to a distorted image of the anatomical structures.

Outline Ultrasound System

The patented Dornier Medizintechnik outline system employs an ultrasound transducer that is mounted on an articulated arm that moves isocentrically in relation to the therapeutic focal point, therefore all movements of the transducer fixed to the articulated arm remain centered at the focal point. This way it is possible to scan the treatment site before and during treatment from different directions, while continuously observing the focal point. Transducers are easily exchanged (5.0 and 7.5 MHz) to achieve optimal adaptation for all treatment situations. The lifetime of the transducer is not adversely affected by shock waves. Because of their direct contact with the body, outline systems offer undisturbed images permitting reliable localization with ultrasound imaging free of artifacts. This system has been applied in millions of patients treated for kidney stones.

Fig. 3.3. Outline ultrasound localization system

Deviations from the sound geometry caused by diffraction are minimal in ESWT because of its subdermal application.

Advantages:
1. All movements remain centered at the focal point.
2. Multidirectional scanning is possible before and during treatment.
3. Standard sectional planes can be selected before and during treatment.
4. Focal point always on the screen represented by crosshairs.
5. Transducer can be easily replaced.
6. The ultrasound device can be used independently for sonography of the musculoskeletal system.
7. Transducer is not exposed to the shock wave field: full lifetime.
8. Direct body contact: undisturbed image free of artifacts: reliable localization.
9. Distortion-free imaging of the anatomical structure using a linear transducer.
10. Variable transducer settings allow either standard diagnostic imaging or imaging of the shock wave path, i.e., the path proximal and distal to the focal point (see Figs. 3.4 and 3.5).

Disadvantages:
1. Osseous structures may be located between the therapy head and the treatment site, thus, obstructing the path of the shock wave.

3.3.2 Isocentric X-ray System

The latest models of ESWT systems have integrated X-ray C-arms allowing for their isocentric movement around the shock wave focus with the help of specially designed soft- and hardware. The focal point always remains visible

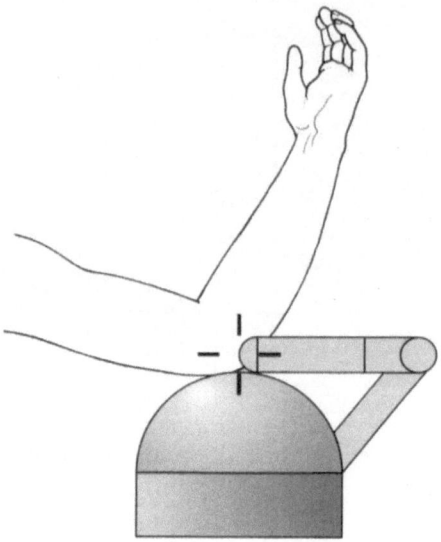

Fig. 3.4. Horizontally coupled transducer

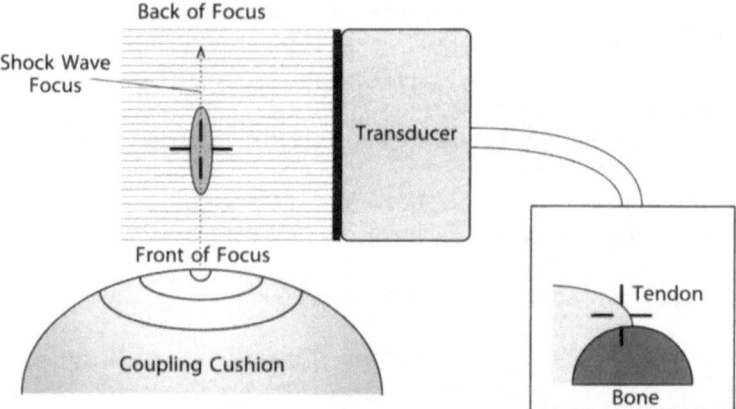

Fig. 3.5. Imaging with horizontally positioned transducer

on the image screen of the X-ray system by tracking with the help of cross-hairs. This permits the adjustment of the focal point to the target by first moving either the shock wave source or the patient in the horizontal plane while performing anterior – posterior fluoroscopy. Then the X-ray device is isocentrically tilted into an oblique position allowing the target to be vertically adjusted. It is recommended that periodic localization checks are made during treatment, especially if the patient is believed to have moved.

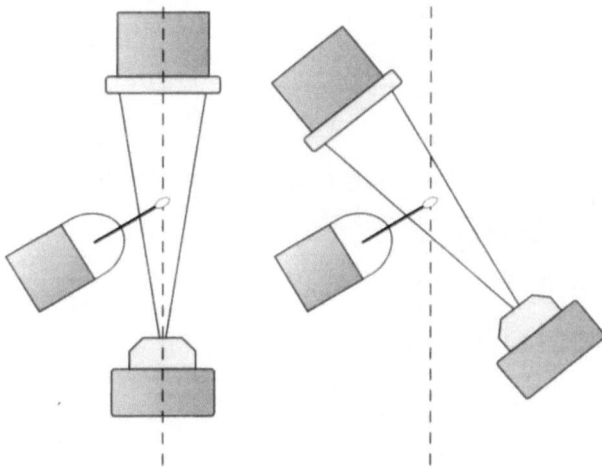

Fig. 3.6. Isocentric X-ray system, rotating around the focal point

---

Advantages:
1. Integration into the shock wave therapy system.
2. All movements are centered at the focal point.
3. Focus always visible on the monitor with the help of crosshairs.

→ Fast and precise localization.

3.3.3 Dual Imaging

The dual imaging technique allows for dual localization. An X-ray system and an ultrasound system are used to follow the course of treatment. This permits precise localization and treatment control, while minimizing the exposure to radiation.

The "Dual Imaging" principle is currently considered to be the best localization method, since both systems meet the multi-directional examination requirements. Both systems can be used for therapeutic and diagnostic purposes simultaneously.

X-ray System and Ultrasound System for Localization
Advantages:
1. Both systems have a multi-directional view of the treatment site.
2. Both systems can be used simultaneously for therapeutic and diagnostic purposes.

→ Precise localization and treatment control with minimal exposure to radiation.

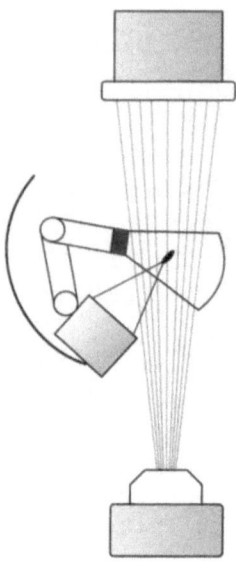

Fig. 3.7. Principles of dual imaging. Simultaneous use of X-ray and ultrasound

3.3.4 Comparison Ultrasound/X-ray Systems

Table 3.1. Comparison of ultrasound and X-ray systems

Sonographic Procedures	Radiologic Procedures
Advantages	
Imaging of soft tissue	Imaging of skeletal structures
Imaging of calcium deposits	Imaging of calcium deposits
No exposure to radiation	Minimal training for interpretation necessary
Simple and easy evaluation	No additional training necessary
Broad public acceptance	
Constant "real-time" control	
Minimal investment	
Safe for the treatment of children	
No special facilities required	
Disadvantages	
Special training necessary	Increased investment cost
Restriction due to adiposity	Potential exposure to radiation
	Requires the use of a separate room

4 Orthopedic Shock Wave Therapy

4.1 Energy Levels

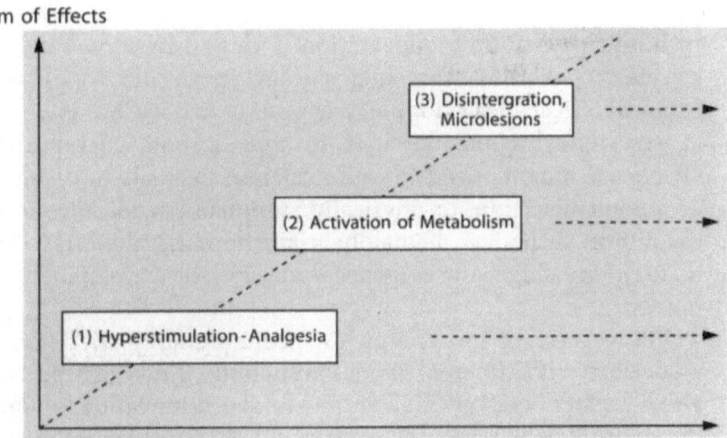

Fig. 4.1. Energy levels

Possible Indications

Table 4.1. Possible indications

Low/Mid Range Energy	Mid/High Range Energy
1. Plantar fasciitis (heel spur)[1]	1. Calcific tendinitis[1]
2. Lateral[1] and medial epicondylitis	2. Plantar fasciitis (heel spur)[1]
3. Supraspinatus syndrome	3. Pseudarthrosis
4. Trochanteric insertional pain	
5. Patellar tendinitis	
6. Achillodynia	

[1] Standard indications with single site, controlled study results existing; multicenter, controlled, double blind study results available soon in the USA and Germany

4.2 Action Mechanism of Shock Wave Therapy

Different shock wave effects have to be distinguished. Their relative importance varies with the energy flux density and the energy levels applied. Generally low, medium and high energy applications are differentiated in orthopedics, according to Rompe[1]. In this book, the associated energy ranges defined with respect to the energy flux density (ED) at the focus are as follows:
Low energy range: 0.08 mJ/mm^2 $<$ ED < 0.28 mJ/mm^2
Mid energy range: 0.28 mJ/mm^2 $\leq$ ED < 0.60 mJ/mm^2
High energy range: 0.60 mJ/mm^2 $\leq$ ED.

The shock wave effects described below are observed.

4.2.1 Structural Damage

In lithotripsy, stone fragmentation is caused by direct and indirect mechanical effects resulting from high energy shock wave application. Indirect mechanical effects lead to the creation of cavitation bubbles. Their collapse causes stone fragmentation. In orthopedic high energy applications, similar effects are hypothesized to cause calcium deposits to disintegrate in the areas of tendon insertion, to physically stimulate pseudarthrotic fracture sites and to initiate secondary hematomas surrounding the sites. The application of high-energy shock waves is necessary for the development of cavitation phenomena.

The success of low-energy to medium-energy shock wave applications is associated with different modes of action, which can be summarized as analgesic effects and effects leading to the stimulation of metabolic reactions (for more details regarding cavitation see chapter 2.11).

4.2.2 Analgesic Effect

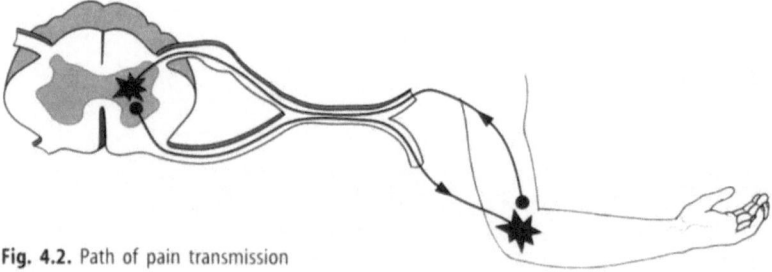

Fig. 4.2. Path of pain transmission

---

[1] Rompe, 1997

Hyperstimulation of axons using stimuli that trigger pain can produce an analgesic effect. This reactive mode of the nervous system, also known as "gate control", is initiated by activating the coreless C-fibers and A-delta-fibers[1].

Signals, transmitted by the core-less C-fibers to the posterior horn of the spinal medulla, reach the periaquaeductal gray matter and are retransmitted back to the posterior horn as inhibitors, causing pain signals to be ignored. Myelinized A-delta-fibers inhibit the transmission of spinal signals from the C-fibers.

Pain memory is lost, normal motion patterns are restored, and neural as well as muscular compensation mechanisms are not required. In this way the vicious circle of pain – central pain registration – pain avoidance by adaptation of compensating motion patterns – development of pathological motion patterns – initiation of pain can be interrupted by readaptive changes of neuronal connections[2].

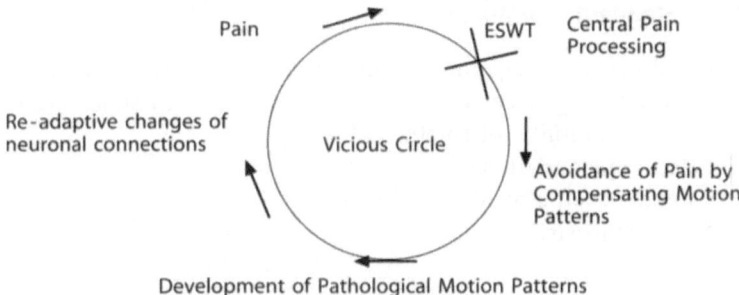

Fig. 4.3. Vicious circle of pain. ESWT interrupts the circle and initiates the healing process

4.2.3 Stimulation of Metabolic Reactions

Shock waves alter the membrane permeability of cells which are in the focal area. Since shock waves do not lead to thermal effects, this effect can be attributed most likely to direct mechanical effects.

Hypothetically pressure waves may change the ionic channels leading to increased intermolecular distances in cell membranes. This way the polarity of nerve membranes is changed causing analgesia by inhibiting depolarization. In addition, metabolic activation may cause stimulation of intra/extracellular ion exchange. Catabolic end products of the metabolism are removed and absorbed. Chronically inflamed tissue is positively stimulated.

Other effects discussed are additional analgesic and antiphlogistic metabolic reactions caused by activation of free radicals (change of the chemical environment and inhibition of neuronal transmitter substances) and macrophage stimulation.

[1] Melzack, 1975

4.2.4 Effects of Shock Waves on the Musculoskeletal System

Table 4.2. Effects of shock waves on the musculoskeletal system

Treatment	Effect	Energy flux density
Bones/pseudarthrosis	Disintegration	High energy (HE)
	Capillary rupture (hematoma)	High energy (HE)
Calcifications	Disintegration	Medium energy (ME)
	Analgesic effect	Low – medium energy (LE – ME)
	Metabolic effect	Low – medium energy (LE – ME)
Tendinitis	Analgesic effect	Low – medium energy (LE – ME)
	Metabolic effect	Low – medium energy (LE – ME)

4.3 Orthopedic Indications for Shock Wave Therapy

4.3.1 Standard Indications

The following orthopedic pathological conditions show statistically significant clinical improvement following treatment with shock waves:
- Calcific tendinitis of rotator cuff
- Lateral epicondylitis
- Plantar fasciitis/plantar heel spur
- Pseudarthrosis.

4.3.2 Clinical Trials

In addition to these standard indications several other orthopedic indications are being investigated either in clinical studies or as empirical therapeutic possibilities with respect to ESWT. In view of the modes of action of shock waves and based on the clinical experience gained so far, ESWT can be assumed to be successful for the indications listed below. The clinical evaluation and 2-year follow-up have not been completed for these indications.
- Dorsal heel spur
- Haglund's exostosis
- Medial epicondylitis
- Supraspinatus tendon syndrome
- Achillodynia
- Patellar tendinitis
- Trochanteric insertional tendinitis.

4.4 Contraindications

Contraindications can be deduced from the modes of action of the shock wave. For example, high mechanical load occurs when the shock wave passes through the interface between tissues of different acoustical impedance. The applied energy dose and the tissue properties forming the interface play an important role.

Shock waves should not be applied directly to large vessels (transition tissue/fluid) due to the risk of potential damage to vascular walls (hemorrhage, inducing thrombosis). The risk of vascular damage or hemorrhage should be considered, when treating MARCUMAR patients. The risk of bleeding is increased in patients undergoing treatment with acetylsalicylic acid (ASA) and heparin.

Large amounts of energy are released when shock waves pass from tissue to air, for instance in pulmonary and intestinal regions. This may lead to ruptured tissue.

During superficial treatment of tendonitis, a hematoma may form at the site of the wave exit (interface tissue/air). It is recommended to apply ultrasound gel to the site of the shock wave exit to protect the skin.

Table 4.3. Contraindications for ESWT

Contraindications	Reason
Disorders of coagulation/ MARCUMAR therapy	Risk of hematoma
Large diameter vessels in focal area	Risk of thrombosis
Organ containing air in the focal area (lung, intestines)	Risk of rupture
Infections in focal area	Risk of activating infection
Tumors in focal area	Risk of activating tumors
Epiphyseal cartilage in focal area	Growth disorder
Areas near the spinal column	Effect on neurons?
Cranial bones in focal area	Effect on neurons?
Pregnancy	Unknown effect
Direct shock wave application to nerves	Change of nerve potential/cellular destruction
Cardiac pacemaker	Interactions

4.5 Patient Information

Patient information should be given in written form. This can be done according to the guidelines for patient information before surgery. It is recommended to hand out patient information material regarding ESWT. Patient information brochures are available. The patient should be advised of the speculative nature regarding the effects of shock waves and about the fact that scientific long-term results are still lacking for ESWT.

The following suggested standard patient consent form does not claim to be complete.

Patient Name:
Name of Physician:
Diagnosis:
Treatment Plan:
Anesthesia (optional):

After a thorough discussion of the nature of my disorder and its expected development, I have been informed of the conservative and surgical treatment possibilities. Treatments conducted since _____ months, such as _____ have not produced the expected results. As an alternative to surgical treatment, which is now indicated, I wish to be treated with extracorporeal shock wave therapy (ESWT). I reject having surgery because of the risks involved with it, because of the status of my disorder and the likelihood of its recurrence.

The procedure for shock wave therapy has been thoroughly explained to me. I am aware of its investigational status, and understand that the long-term clinical results of ESWT are unknown. I also understand the therapeutic failure rate is estimated to be between 20% and 30%.

After an indepth discussion, which I fully understood, I had the opportunity to ask questions to which I received complete and comprehensive answers. I understand that I can ask the attending physician additional questions at any time. It was made expressively clear to me that the success of the procedure cannot be guaranteed.

Potential risks and side effects of shock wave therapy, such as injury to bones, vessels, nerves and hematoma formation, were explained to me in detail.

I was informed of pre- and post-therapy treatment requirements and of the necessity for follow-up examinations.

I assure that I have fully disclosed all previous illnesses, disorders and abnormalities to the treating physician.

Having been informed of all treatment alternatives and after failure of prior treatments, I am willing to undergo extracorporeal shock wave therapy (ESWT) as another form of treatment.

Signature Patient   Signature Physician

_____   _____

_____ Date   _____ Date

Fig. 4.4. Standard patient consent form

5 Localization Techniques

5.1 Patient Feedback

When locating the ROI (region of interest) with the help of patient feedback, the point of maximum pain is palpated and marked (X) on the skin. Using the marking of the shock wave path on the coupling cushion, the therapeutic focus is placed either directly (OPM = one point method) or tangentially (TPM = two point method) at the ROI.

For tangential coupling: Identify the coupling point (O) and verify that the laser pointer and tender point (X) are superimposed.

For direct coupling: Verify that therapy head center point and tender point (X) are superimposed.

The laterally attached laser pointer marks the projection of the geometric focal point on the patient's skin. Filling or draining the coupling cushion determines the penetration depth of the therapeutic focal point (minimal water pressure = maximum penetration depth; maximum water pressure = minimum penetration depth).

Fig. 5.1. Laser pointer and marking of the shock wave path on the coupling cushion

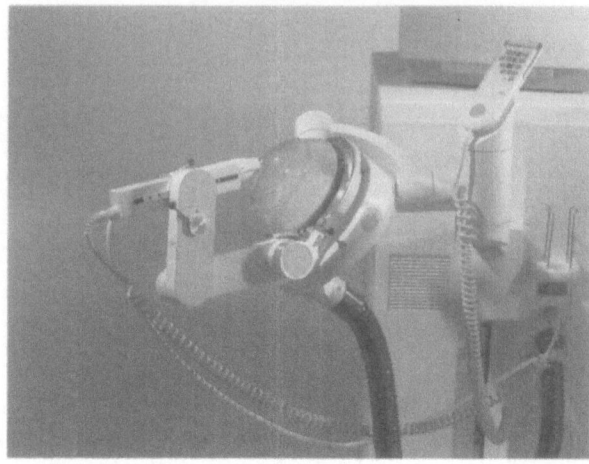

Fig. 5.2. Isocentric outline ultrasound localization system

5.2 Ultrasound

Isocentric outline ultrasound localization permits precise identification of the treatment area (ROI – region of interest).

In contrast to inline ultrasound localization, with outline ultrasound localization it is possible to perform localization from various angles without interference from artifacts and thus to achieve more precise localization. The transducer can be maneuvered into the optimal position by rotation around the therapy head and, thus, around the focal point. The transducer can be rotated for localization of the ROI in two planes. Identification of anatomical structures is greatly simplified by the maneuverability of the system. The area anterior and posterior to the therapeutic focus can also be observed.

For all standard localization procedures, at least two DEGUM (DE-GUM = German Society for Ultrasound Medicine) required standard sectional planes can be imaged.

5.3 X-ray

The latest models of the ESWT systems have integrated X-ray C-arms that permit isocentric movement of the C-arm around the shock wave focus with the help of specially designed soft- and hardware. The focal point always remains visible on the image screen of the X-ray system by tracking with the help of crosshairs. This design permits the adjustment of the focal point to the target by first moving either the shock wave source or the patient in the horizontal plane while performing anterior-posterior fluoroscopy. After this, the X-ray device is isocentrically tilted into an oblique position allowing the target to be adjusted vertically. Periodic localization checks are recommended during treatment, especially after the patient has moved.

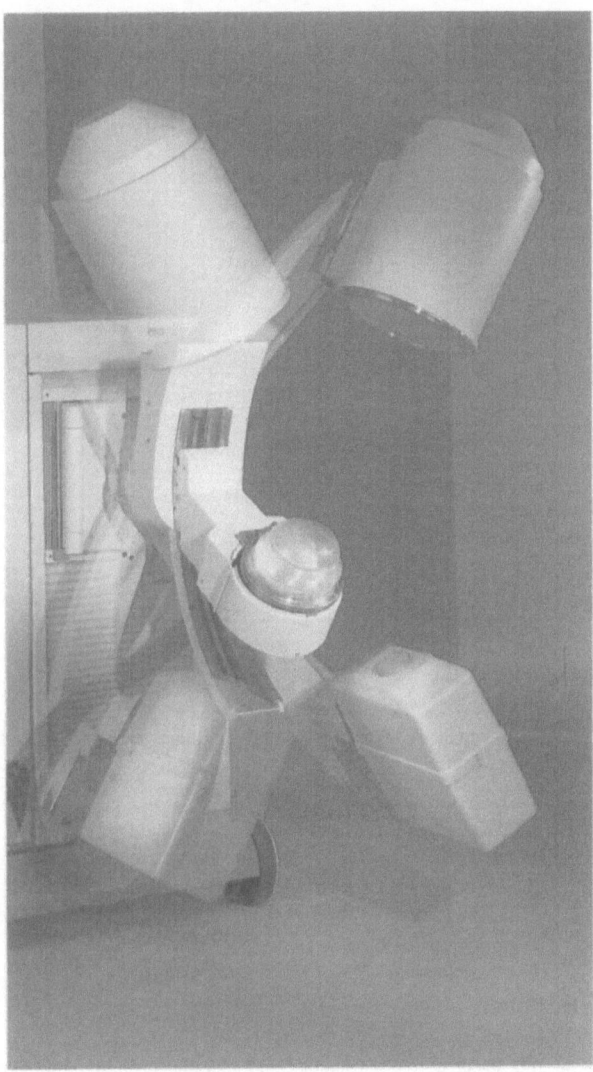

Fig. 5.3. Integrated rotating isocentric X-ray C-arm

5.4 Dual Imaging

Dual imaging is achieved by independent localization with two systems. Following the use of the X-ray system, the ultrasound system is used to follow up the course of treatment. This technique insures both precise localization and treatment control with minimal exposure to radiation.

When the dual imaging method is employed, both systems meet the requirement that the treatment site can be observed in all spatial directions. Both systems can be used for diagnosis and control during the treatment.

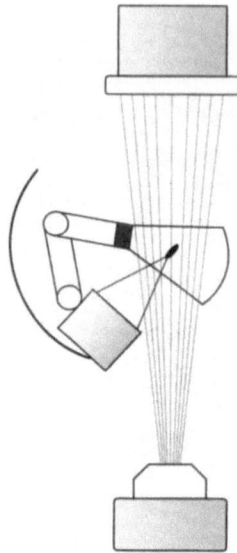

Fig. 5.4. Principles of dual imaging

5.5 Comparison of Imaging Procedures

X-ray

Table 5.1. Advantages, disadvantages and indications for X-ray localization

Advantages	Disadvantages	Indications
1. Localization of bones/ calcium deposits	1. Exposure to radiation 2. Static image 3. Does not image soft tissue 4. Purchase price relatively high	1. Pseudarthrosis 2. Calcific tendinitis 3. Plantar fasciitis

Ultrasound

Table 5.2. Advantages, disadvantages and indications of ultrasound localization

Advantages	Disadvantages	Indication
1. Diagnostic imaging and localization of soft tissue structures and its pathological changes (calcium deposits) 2. Real-time imaging with permanent control of the focal point 3. No exposure to radiation 4. Safe for use with children 5. Multifunctional use of device in doctor's offices 6. General acceptance by patients 7. Relatively good cost-benefit relationship	1. Not suitable for pseudarthrosis in case of large bones 2. Requires special training	1. Lateral and medial epicondylitis 2. Calcific tendinitis 3. Supraspinatus tendinitis 4. Trochanteric tendinitis 5. Patellar tendinitis 6. Achillodynia 7. Plantar fasciitis

6 Application Techniques

6.1 Tangential Coupling

Tangential shock wave direction should be selected whenever possible. This allows

- optimal positioning of the shock wave focus with respect to its longitudinal and lateral dimensions
- precise localization of the insertion plane at the point of bone/ligament insertion
- positioning of transducer directly above the ROI when using ultrasound localization
- minimization of the risk of injury to surrounding structures (e.g., lungs when treating shoulder)
- avoidance of periostal over-stimulation, which can often trigger pain response.

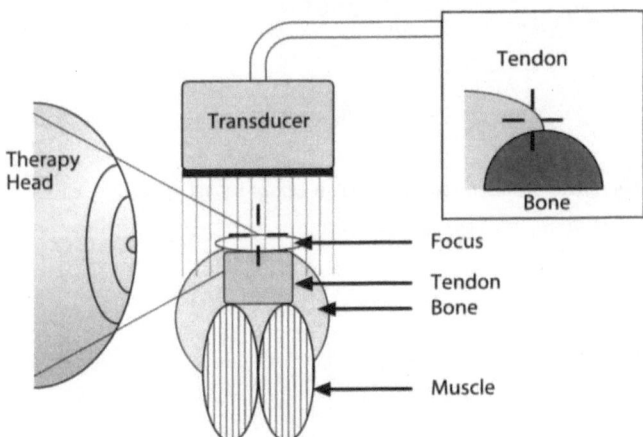

Fig. 6.1. Tangential coupling of the therapy head

6.1.1 Two Point Method (TPM)

Both, laser pointer and ultrasound localization can be achieved quickly and reliably using this method:

Step 1: Localization and marking of tender point (X)

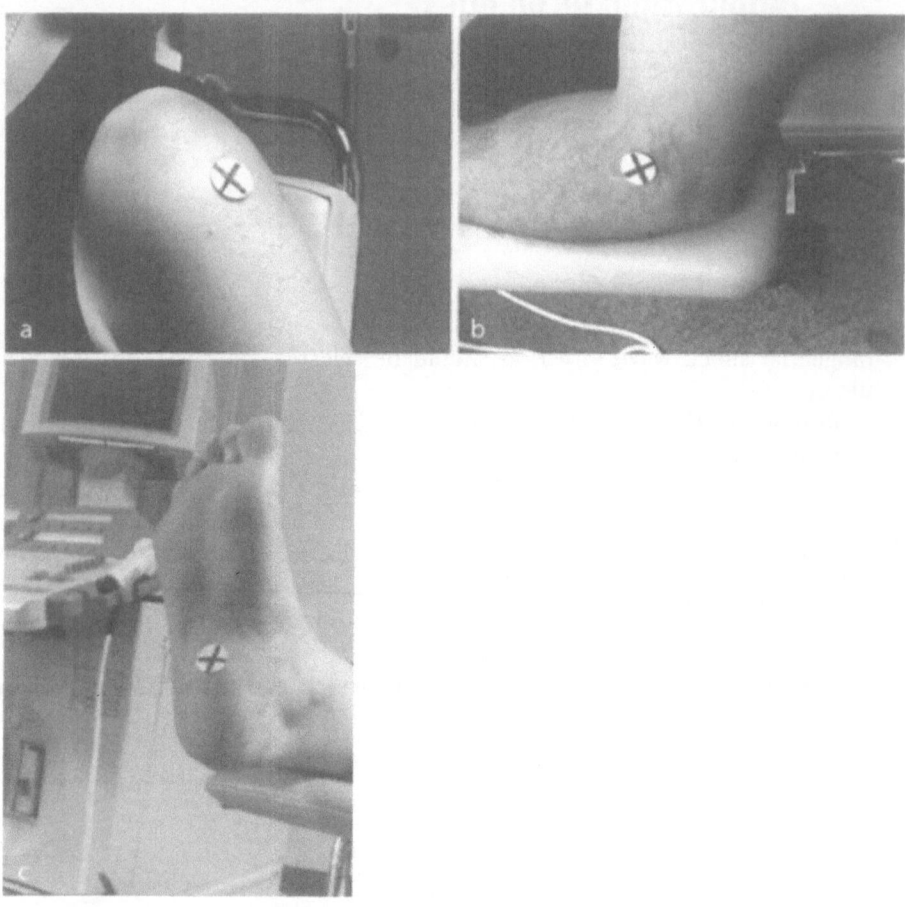

Fig. 6.2. a Shoulder: Lateral contour. **b** Elbow: Lateral epicondyle. **c** Heel: Medial-plantar calcaneus

Step 2: Localization and marking of coupling point of therapy head (O)

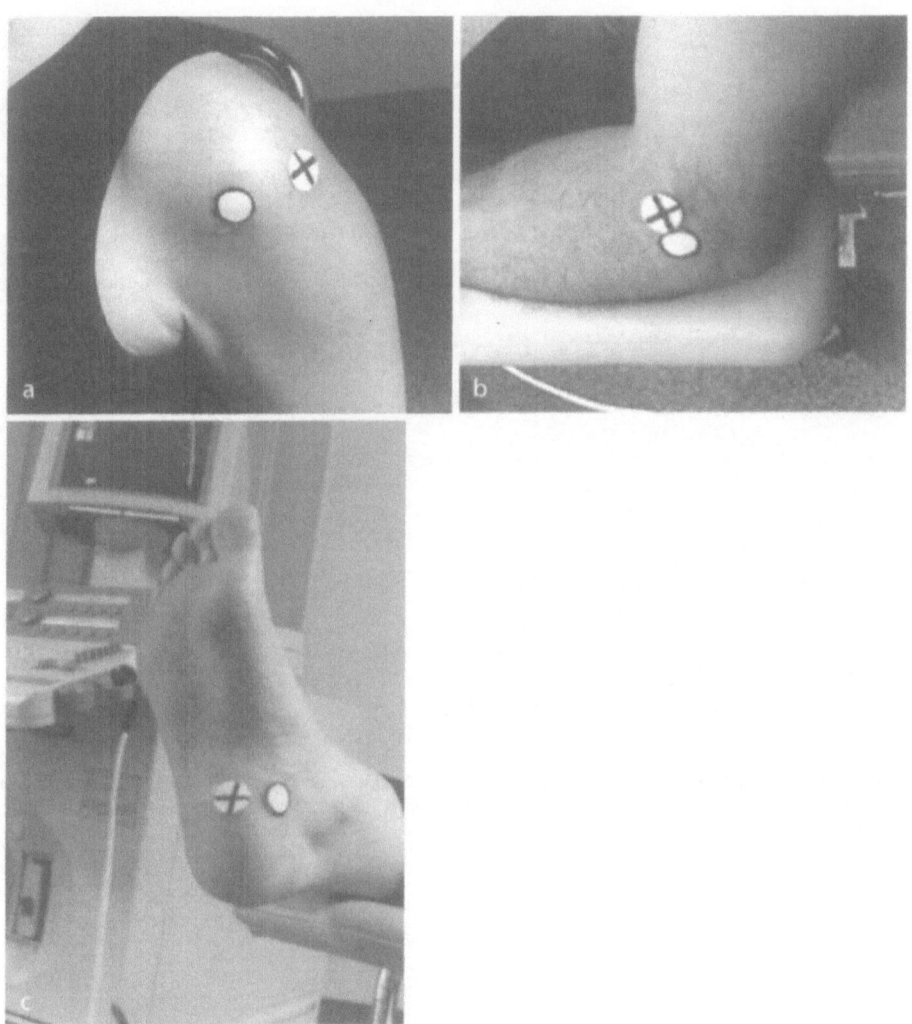

Fig. 6.3. a Shoulder: Ventral coupling. **b** Elbow: Lateral coupling. **c** Foot: Medial coupling

Step 3: Placement of therapy head (O); tender point (X) = center of focal point

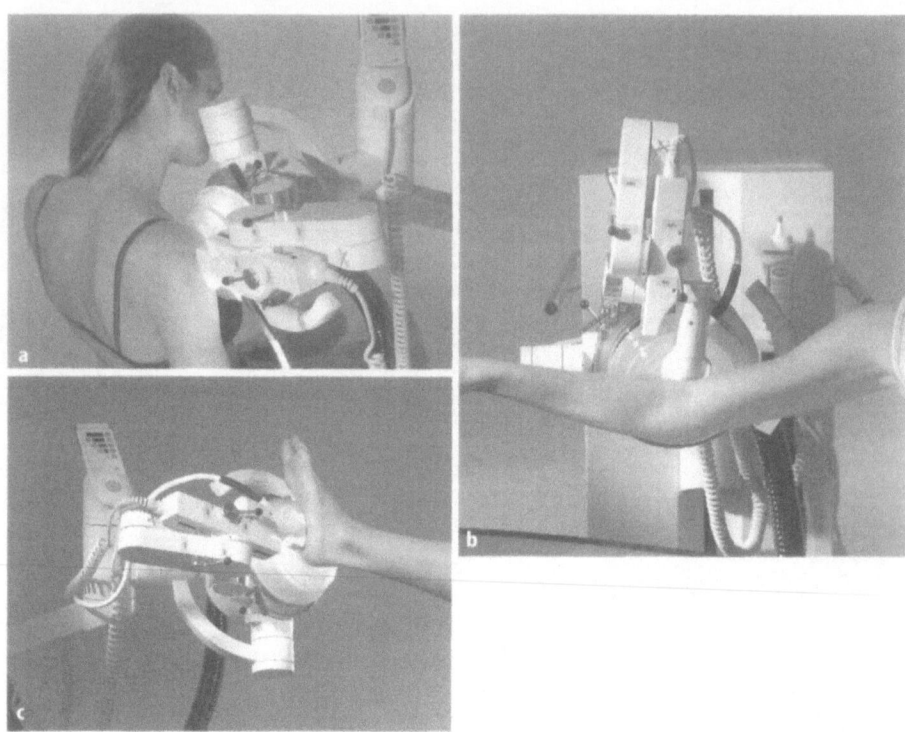

Fig. 6.4a–c. Tangential coupling: Adjust laser pointer or ultrasound transducer to tender point (X) as shown in Fig. 6.2a. **a** Tangential coupling to the shoulder. **b** Tangential coupling to the elbow. **c** Tangential coupling to the foot

6.2 Direct Coupling

For this type of coupling, the focus is applied vertically to the ROI (X). The ROI or the tendon insertion area respectively is treated.

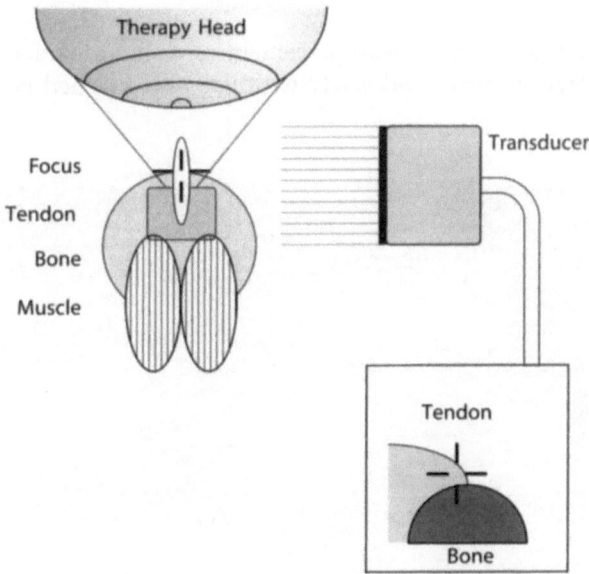

Fig. 6.5. Direct coupling of the therapy head

6.2.1 One Point Method (OPM)

The tender point (X) and the coupling point (O) are at the same location for this method (therefore: one point method).

Step 1: Localization and marking of tender point (X).

Step 2: Therapy head placement directly on tender point (O). Adjust the penetration depth and verify with laterally attached laser pointer.

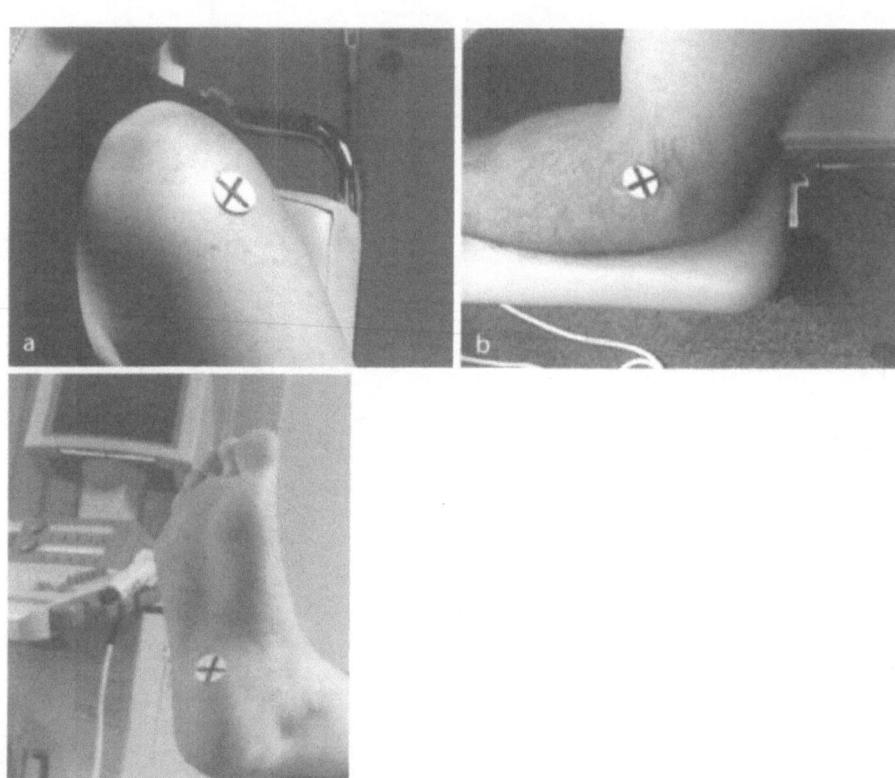

Fig. 6.6. a Lateral contour of shoulder. **b** Lateral epicondyle. **c** Heel: medial-plantar calcaneus

7 Ultrasound Examination – Standard Ultrasound Cross Sectional Planes (DEGUM Recommendations)

7.1 Introduction

Ultrasound examination of the joints of the musculoskeletal system can supply important additional information for treating a number of clinical conditions and injuries. This can apply to both intraarticular structures, such as hyaline cartilage and periarticular soft tissue structures seen in calcific tendinitis, or rotator cuff defects. Especially in the case of degenerative and inflammatory changes of the joints, the availability of ultrasound cross-sectional images makes a valuable contribution. The presented standard cross-sectional planes are oriented according to leading osseous structures. For the physician with limited ultrasound experience, these structures enhance the initial learning curve. For experienced personnel, these standard cross-sectional planes facilitate quick and easy adjustment of the transducer, maximizing the sonographical overview of the investigated joint. Beyond this the documented images of standard cross-sectional planes are easily interpreted by other experienced personnel. This chapter describes the basic ultrasound techniques and standard ultrasound sectional planes of presented joints using DEGUM guidelines (Meeting of the work group Musculoskeletal System, January 20, 1996).

7.2 Safety Aspects, Advantages and Disadvantages of Ultrasound Examinations

Ultrasound examinations are radiation free and are relatively risk-free for patients and medical personnel according to the present state of knowledge. The EFSUMB (European Federation for Security of Ultrasound in Medicine and Biology) has established a commission (The Watch Dog Group) with the objective to monitor potential side effects of ultrasound devices applying continuously improved testing methods. According to the American Institute for Ultrasound in Medicine (AIUM) (October 1978, last revision October 1982) in the frequency range of a few MHz no adverse effects have been reported on mammal tissues which were exposed to less than 100 mW/cm^2 up to now. Likewise, there were no side effects detected at higher intensities with exposure times between 1 and 500 seconds, when the product of sound intensity and exposure time was less than 50 J/cm^2.

Commercially available ultrasound devices have ultrasound intensities in the range of less than 100 mW/cm^2 , and therefore are approved for use without limitation on treatment time.

Advantages
1. Relatively risk-free for the patient or the healthcare professional.
2. No exposure of patient or healthcare professional to radiation.
3. Online ultrasound examination allows visualization of the effected joint in motion facilitating identification of pathologies.
4. Enables bilateral comparative examination.
5. Treatment can be monitored, when metallic total joint replacements are present.
6. Low cost.

Disadvantages
1. Only those cortical structures that are facing the transducer can be imaged by ultrasound.
2. Cortical structures facing away from the transducer cannot be examined by ultrasound.
3. Intra-osseous changes cannot be examined by ultrasound.

Following patient history and physical examination, the sonographic examinations should be conducted before initiating X-ray examinations.

Please note: Biplane radiological examination cannot be replaced by sonography.

7.3 Technical and Physical Principles of Ultrasound Examinations

The principle of ultrasound examination is based on the piezoelectric effect and its inversion. When pressure or tension is exerted on certain crystals (generally quartze or tourmaline), this induces a surface charge leading to an electric field applied to the crystal (the piezoelectric effect). On the other hand if a voltage is applied to piezoelectric crystal, this causes the crystal to change in size (the inverse piezoelectric effect). The same effects occur when alternating pressures or voltages are exerted on the piezoelectric crystal inducing alternating voltages or volume changes of the crystal, respectively. The latter effect can be used for the emission of sound waves, while the former allows the receiving and analyzing of ultrasound waves, since ultrasound waves exert alternating pressure on the crystal. An electromechanical transducer (ultrasound transducer) emits and receives ultrasound waves applying these principles. This technique is called the impulse-echo method. A sound wave (impulse) passing through tissues triggers an echo, which can return to the ultrasound transducer, if the transducer is correctly positioned.

Sound wave propagation can be described similar to the propagation of light in wave optics. In homogeneous tissue sound waves propagate in straight lines. The number of vibrations (frequency) per second is expressed in Hertz (Hz). The range, which is audible to the human ear, is between 16 and a maximum of 20,000 Hz. The range below 16 Hz is called infrasound; the range above 20,000 Hz is called ultrasound. Diagnostic frequencies used at present for sonography of the musculoskeletal system range from 5 to 12 MHz, and therefore are not audible. Sound wave velocity depends on the material in which it travels: In air the sound wave velocity is 300 m/s; it increases with increasing density of the material. In water, sound travels at a velocity of 1,497 m/s, in the human body sound velocity ranges from 1,490 and 1,610 m/s. Diagnostic ultrasound wave lengths vary between 1.5 mm and 0.128 mm. In the range of 20–37 °C, sound velocity is practically independent of temperature variations. The reflection of a sound wave at an tissue interface depends on the acoustic difference of the tissue involved and the angle of incidence of the impacting wave. The most important acoustic tissue properties are density and elasticity. These parameters determine the speed of sound and the acoustical impedance of the tissue. Acoustic impedance corresponds to sound resistance. The greater the difference in acoustical impedance between two imaged structures, the easier it is to differentiate them sonographically. As sound waves propagate, they are influenced by refraction, diffraction, scattering, reflection and absorption in various ways.

Refraction: When passing from one medium to another, ultrasound waves are refracted at the boundaries according to the difference in acoustic impedance. Ultrasound waves with vertical incidence do not change their direction of propagation when passing from one medium to another. Thus, when the returning echoes hit the interfaces between different media at right angles or nearly right angles, the transducer receives the returning ultrasound waves with less attenuation effects than the incoming waves with oblique incidence.

Diffraction: When sound waves that normally travel in a straight line through a homogenous medium encounter an obstacle, they are diffracted and a sound shadow occurs distal to the obstacle. The sound intensity is reduced distally. Diffraction decreases as sound frequency increases.

Scattering: Because interfaces in the human body are generally rough, ultrasound waves are scattered at these interfaces. This leads to a loss of energy. Scattering is directly proportional to the frequency.

Absorption: Ultrasound wave energy is weakened, depending on the tissue type it passes through. Tissues absorb energy and convert it to heat. To compensate for this loss of energy, images of deep tissue structures must be amplified. This is accomplished with the aid of time gain compensation controls (see time gain compensation controls). These controls allow amplification of deep tissue structures depending on time of propagation of the ultrasound beam.

7.4 Configuration of Ultrasound Systems

Ultrasound systems consist of the ultrasound device, a monitor and at least one ultrasound transducer. The ultrasound device contains electronics for signal processing. The transducer receives signals from the area of examination. These are transformed into up to 256 different shades of gray. Since the human eye cannot differentiate between this many shades of gray, the ultrasound device reduces them to a scale of approximately 20 shades, which then appear on the monitor. The transducer is simultaneously an emitter and a receiver of sound waves. Every transducer has a near field, a mid field and a far field (Fig. 7.1).

The near field is immediately proximal to the transducer. The mid field is the region of highest resolution, where structures of interest are to be imaged. The mid field lies between the near field and the far field region of the transducer. Structures in the near and far field regions of the transducer cannot be safely evaluated by ultrasound. The distant or far field region is the region furthest away from the transducer. In this region sound shadows and repetition artifacts can be identified. The transducer is placed directly on the skin covering the area of examination. A contact gel, which consists essentially of gelatinous, highly viscous water, must be used to cover the skin. This is necessary, because air leads to total reflection of sound waves. Only by using contact gel is it possible to see an image on the monitor. In the event the transducer and the skin surface are not congruent (maximally flexed knee joint, for example) or in cachectic patients with prominent bony structures, it may be necessary to use a stand-off to obtain a smooth surface on which to place the transducer. The use of a stand-off increases the distance to the near field, the mid field and the far field of the transducer.

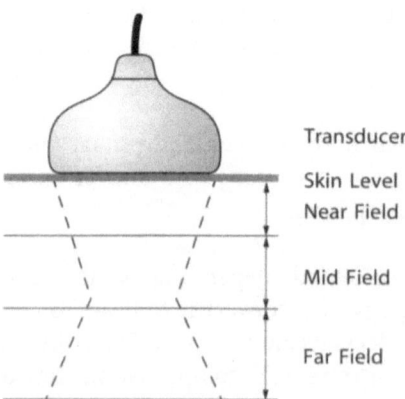

Transducer

Skin Level
Near Field

Mid Field

Far Field

Fig. 7.1. Near field, mid field and far field of the ultrasound transducer

Therefore penetration depth may be insufficient, when using a transducer with a high frequency. A sound wave frequency range of 5 to 12 MHz is most commonly used for imaging of musculoskeletal systems today. The sound frequency is directly proportional to the resolution and inversely proportional to the penetration depth. Principally, transducers with the highest possible frequency should be used. In a so-called real-time ultrasound examination, the area of examination is imaged directly on the monitor as a moving picture in shades of gray.

7.5 Transducers, Imaging

Three different types of imaging methods are employed in the B-mode (brightness modulation) technique: compound scan, sector scan and linear scan. The last two are the preferred methods. Compound scans produce a snapshot, while sector and linear scans permit direct visualization in real time. The linear transducer is currently the transducer of choice for ultrasound imaging of the musculoskeletal system. Sector transducers are presently indicated solely for meniscus sonography and for ultrasound examination of the anterior glenoid labrum in the case of Bankart's lesion, an osteochondral shearing fracture, resulting most frequently from an anterior shoulder joint subluxation (Fig. 7.2).

Linear Transducer: The transducer surface is planar. Quartz elements on the transducer surface have a linear configuration. The monitor shows a full-screen square image. By electronic focusing, individual transducer elements

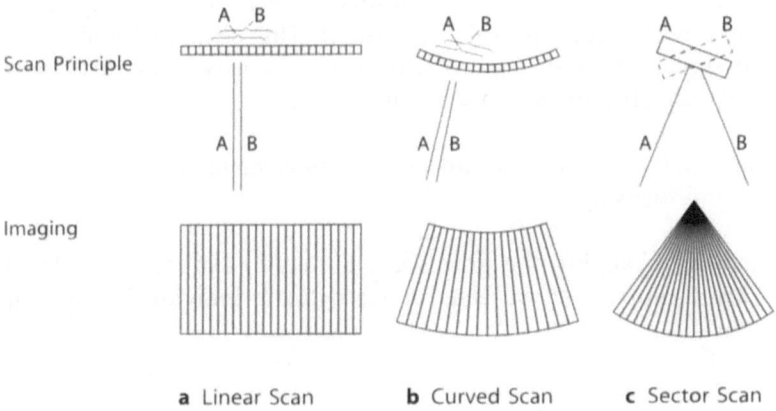

Fig. 7.2. Scanning principles and imaging. Linear transducer, curved-array transducer and sector transducer (from: R. Graf and P. Schuler: Sonography of the Musculoskeletal Systems in Adults and Children, 2nd ed., Chapman & Hall, London, Glasgow, Weinheim, New York, Tokyo, Melbourne, Madras, 1995)

can be triggered selectively. In this manner regions of varying imaging depths can be focused selectively.

Curved Array Transducer: The transducer surface has a moderately convex curvature. The quartz elements on the transducer surface show a convex arrangement. The resulting divergent sound beam produces a trapeze-shaped image on the monitor, which does not fill out the screen.

Sector Transducer: The transducer surface has a highly convex curvature. The quartz elements on the transducer surface are arranged in sectors and rotate around a stationary point with a fixed axis of rotation. The ultrasound impulse is emitted in various directions. On the monitor this results in a pie shaped conical image, which does not fill the monitor screen. Sector transducers may be mechanically or electronically controlled.

Lateral Resolution: This refers to the minimum distance between two objects perpendicular to the sound beam direction still to be discriminated. Lateral resolution corresponds to 4–5 wavelengths.

Axial Resolution: This refers to the minimum distance between two objects in the direction of the sound beam still to be discriminated. Axial resolution corresponds to 2–3 wavelengths.

Lateral and axial resolution are dependent upon sound frequency, pulse length and width of the sound beam.

Time Gain Compensation (TGC) Controls: Sound waves lose intensity while propagating. The highest sound intensity exists directly at the transducer. To receive signals from deeper lying parts of the examined region these signals must be amplified. It is recommended that minimum TGC is not or just slightly activated for regions near the transducer, while it is linearly amplified with increasing distance from the transducer. This leads to amplification of ultrasound signals from all regions with equal intensity, thus, also allowing imaging of deep structures on the monitor.

Pre-processing: Electronic delay and integral action elements serve to improve signal quality and focusing.

Post-processing: By changing the gray scale, a stored (frozen) sonogram is corrected. This action either weakens or strengthens individual gray scale areas.

7.6 Stand-offs

Stand-offs were previously used in ultrasound examinations of the musculo-skeletal systems to visualize subcutaneous structures (e.g., Achilles tendon, supraspinatus tendon) when using 5 MHz transducers. This shifts proximal anatomical structures away from the near field zone, which cannot be evaluated, into the mid field, or focal zone, of the transducer. Stand-offs also allow the transducer to be placed on a curved surface, such as the knee joint, when maximally flexed, without producing coupling artifacts. With today's 7.5 MHz linear transducers, such stand-offs are no longer required. They are, however, frequently required for dynamic examinations of the Achilles tendon in plantar flexion and dorsal extension of the foot. All stand-offs create additional artifacts. For use on the musculoskeletal system the ideal stand-off thickness is between 0.5 and 2 cm. The stand-off should be cleaned with warm water after use. The use of water-filled surgical gloves as stand-offs is not recommended. Solid body and water stand-offs are recommended for the applications presented here.

Solid body stand-off (e.g., Proxon®): Highly recommended, easily maintained. So-called "softeners" are added to these solid bodies. These "softeners" can soften plastics when allowed to act on plastic surfaces for an extended period. Thus, they should not be placed on the plastic surfaces of ultrasound devices or similar products for extended periods.

Water stand-off: Water stand-offs are intensive in maintenance and require a weekly exchange of the water. Therefore, they are recommended with reservation. Tap water contains suspended particles, which affect image quality. It should not be used. The important advantage of a water stand-off compared to a solid body stand-off is its easy adaptation to the shape of the transducer. This permits the transducer and stand-off to be placed in one hand, while the other hand remains free to perform the examination.

Gel stand-off: Gel or agar stand-offs are not recommended due to their limited life times.

7.7 Artifacts and Phenomena

Artifacts are artificial products having no or little relationship to the object being examined. Phenomena are peculiarities of the image, which are related to the object under examination. In the following the most common artifacts and phenomena are listed.

Repetition Artifact: Repetition or reverberation artifacts are caused by "caught" sound waves bouncing back and forth between two or more parallel interfaces. The multiple reflection of sound waves leads to increasingly weaker echo impulses returning to the transducer. A typical example of a repetition artifact can be observed when a water cushion is used: The membrane of the rubber foil that surrounds the water is imaged as parallel, echogenic lines on the monitor (Fig. 7.3).

Coupling Artifact: Coupling artifacts occur when an air pocket is formed between the patient's skin and the transducer surface. Usually this is caused by an insufficient amount of gel. Another example of insufficient coupling occurs when the transducer is placed on a convex structure (e.g., knee joint ventral, when maximally flexed) (Fig. 7.4).

Arch-shaped Artifact: An arch-shaped artifact appears, when a highly reflective structure is imaged in the echo-free region. For example, when imaging a wire in a water bath in the transverse sectional plane, so-called "secondary clubs" appear echogenic at both sides (Fig. 7.5).

Phenomenon of a Pseudo-lesion: A pseudo-lesion is observed when curved osseous surfaces are not imaged as end-to-end cortical structures due to oblique incidence of the sound beam. This is observed while scanning the femoral condyle, humeral trochlea and humeral capitulum, etc. This problem

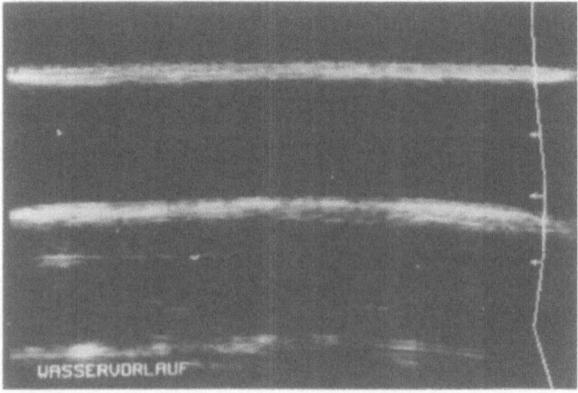

Fig. 7.3. Repetition artifacts with use of water stand-offs

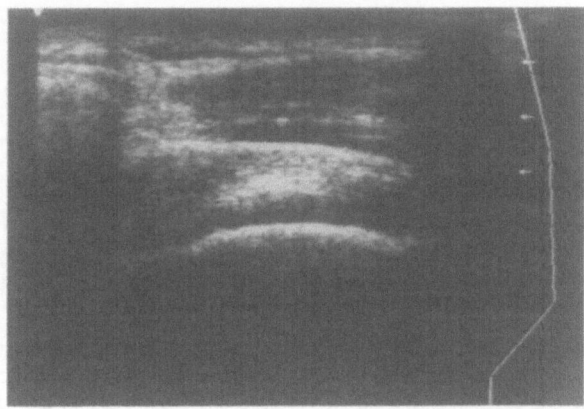

Fig. 7.4. Coupling artifact (shoulder joint: longitudinal plane, lateral-superior region). An echo-free zone develops when the transducer is not correctly placed on the skin (right edge of image)

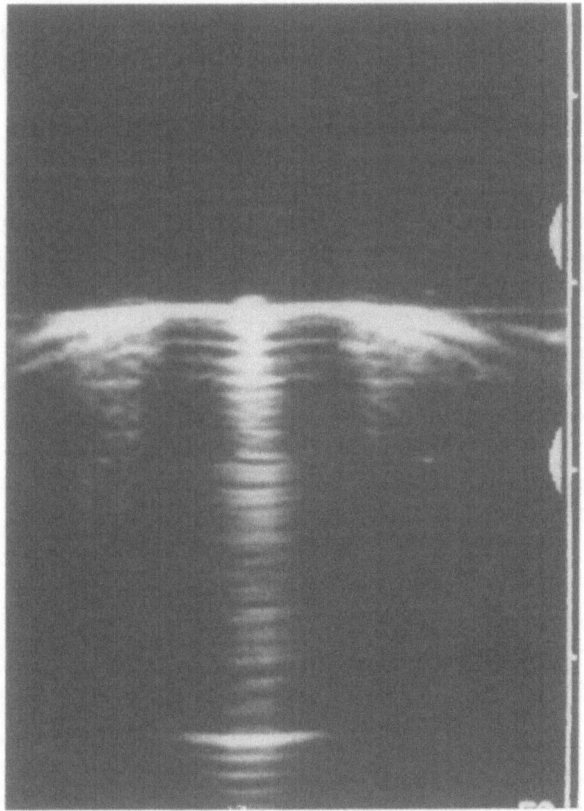

Fig. 7.5. Arched artifact: The cross-sectional plane of a wire in a water bath does not appear as a dot shaped structure but, due to its "side clubs", as an umbrella shaped structure in the ultrasound image

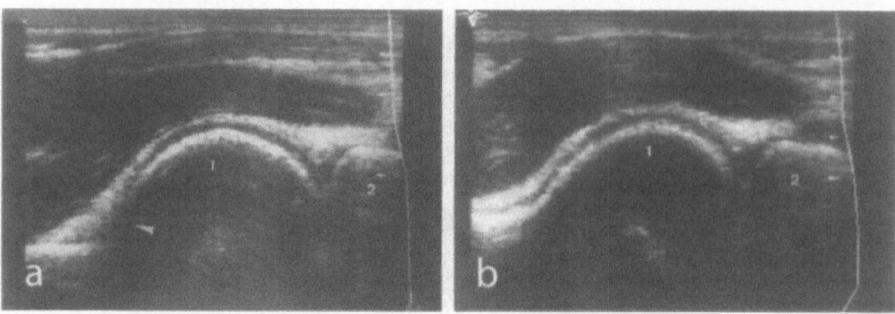

Fig. 7.6 a, b. Phenomenon of pseudo-lesion. Elbow joint, ventral humero radial longitudinal sectional plane. **a** Pseudo-lesion (arrow) in the area of the humeral capitulum. **b** Previously fractured cortical structure appears as one piece, when the direction of the sound beam is slightly altered. 1 Humeral capitulum; 2 Radial head

is caused by poor imaging of the physiologic curvatures for physical reasons. It is possible to distinguish this phenomenon from a lesion:

- In the case of a pseudo-lesion, no basis reflex is present contrary to the case of a lesion.
- The osseous region can be imaged by ultrasound as a single anatomical structure (Fig. 7.6a and b) by either changing the direction of sound or by modifying the position of the joint.

Phenomenon of Reflex Reversal – Phenomenon of the Wandering Reflex (also commonly known as anisotropy): This is another phenomenon observed in tendon tissue. When tendon tissue is sonographically examined it appears either echo poor or echogenic depending on the angle of incidence of the incoming ultrasound. The portion of the tendon that is impacted at a right angle appears physiologically as echogenic (hyperechoic). The portion, which is not impacted by ultrasound waves at a right angle appears physiologically as echo poor (hypoechoic). This may be attributable to an oblique angle of incidence of sound waves, or to a convex or concave structure of the tendon. If the tendon is curved, the image of its structure will change from an echogenic (hyperechoic) to an echo poor (hypoechoic) image. This is called the phenomenon of a wandering reflex.

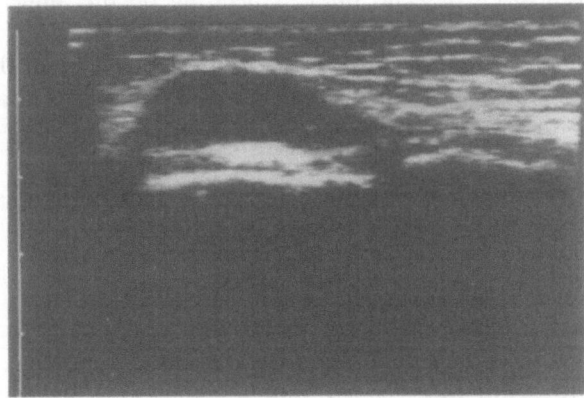

Fig. 7.7. Burning glass effect (enhancement), (knee joint, medial longitudinal sectional plane): Soft tissue structures located immediately lateral to the medial meniscus ganglion are imaged hyperechoic compared to adjoining soft tissue structures

7.8 Important Basic Ultrasound Terms

Burning Glass Effect: Also called "sound intensification" or "enhancement", occurs in ultrasound examinations of liquid-filled hollow spaces, e.g., vessels, Baker cysts or meniscus ganglion (Fig. 7.7). Measured relative to neighboring soft tissue areas, more sound energy is available dorsal to this liquid-filled soft tissue area due to reduced sound absorption. This leads to hyperechoic presentations on the monitor of structures, which appear dorsal to the liquid-filled region (like sound magnification by liquids).

Sound Shadow: If the sound beam hits a structure impenetrable to sound waves, an echo-free zone appears behind it (i.e., distal to the transducer, Figs. 7.8a and b). The imaged structure creating the sound shadow may be

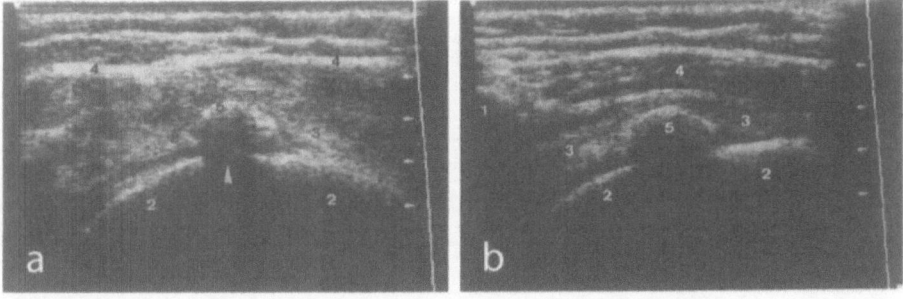

Fig. 7.8a and b. Sound shadow, shoulder joint, lateral-superior transverse sectional plane **a** and longitudinal sectional plane **b**. In calcific tendinitis, the calcium shadow appears as an echogenic structure with a sound shadow (arrow) in both planes. 1 Acromion; 2 Humeral head; 3 Supraspinatus tendon; 4 M. deltoid; 5 Calcium deposit

echo rich or echo poor, depending on the angle (perpendicular or oblique) of incidence.

As an indication for a calcium deposit, often the examiner recognizes the sound "extinction" in the area of the humerus cortex layer first, and then the echogenic change in structure in the tendon area.

Sound Extinction: If an acoustically impenetrable structure, e.g., bone, is underneath a sound shadow, sound extinction will occur within the area covered by the sound shadow.

Reflection: Portions of the sound wave are reflected from the interface between two media with different sound velocities. If structures are impacted by sound waves at an oblique angle, arriving sound waves are reflected depending upon the angle of incidence (angle of incidence = reflecting angle) and only part of the sound waves will reach the transducer.

7.9 Special System Adjustments and Problems of Ultrasound Examinations of the Musculoskeletal System

Osseous structures in sonography of the musculoskeletal system serve as guiding structures for proper alignment of the standard sectional planes. However, they also pose obstacles, because areas immediately posterior to the skeletal anatomy are hidden from the sonographic view. The essential advantage of sonography compared to other sectional imaging techniques is the capability for dynamic imaging. At times structural functional disorders can only be recognized by this technique. Another advantage is the ability for bilateral comparison, which should be used under all circumstances. For certain standard sectional planes the coupling of the transducer to convex formed parts of a joint may benefit from the use of a stand-off. The most frequently used stand-offs are made out of Proxon, gel or water all having their specific advantages and disadvantages (see above).

7.10 DEGUM Guidelines for Documentation and Evaluation

For ultrasound examinations of the musculoskeletal system it is of great advantage to follow a standardized evaluation procedure like for other imaging techniques. In German speaking countries, imaging is governed by the guidelines of the work group Musculoskeletal System of DEGUM (Table 7.1).

Table 7.1. Monitor image and documentation of findings according to the guidelines of the work group Musculoskeletal System of DEGUM

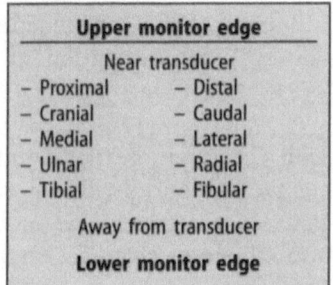

Left monitor edge

	Upper monitor edge	
	Near transducer	
	– Proximal	– Distal
	– Cranial	– Caudal
	– Medial	– Lateral
	– Ulnar	– Radial
	– Tibial	– Fibular
	Away from transducer	
	Lower monitor edge	

Right monitor edge

Ultrasound examination of hip joints of infants is an exception, with the patient's proximal anatomy being imaged on the right side of the monitor screen. Imaging must always be "white on black", i.e., osseous structures must be imaged as white, echogenic structures against black background. In comparison to osseous structures, soft tissue appears as an echo poor image. It is recommended to evaluate anatomical structures visualized in the ultrasound image in a standardized sequence:
- Osseous structures
- Joint cavity and bursa
- Tendons, muscles and periarticular soft tissue.

Changes in shape and echogenic properties of imaged structures are evaluated using dynamic imaging.

Evaluation Criteria for Osseous structures
- Change of shape above level (e.g., osteophyte, exostosis)
- Change of shape below level (e.g., lesion)
- Weakening of reflective pattern
- Reinforcement of reflective pattern.

Evaluation Criteria for Joint Cavity and Bursa
- Usually imaged as narrow bands with dense echoes
- Change of shape in presence of intraarticular volume increase.

Evaluation Criteria for Tendons, Muscles and Periarticular Soft Tissue Structures
- Changes in the shape or absence of anatomical structures are reliable pathological indicators.

Ultrasound examinations do not allow for classification based on histological criteria!

Image description: Should be performed according to standardized terms used in sonography (echo free, echo poor echogenic – homogenous – not homogenous).

Changes: May involve shape, echogenic behavior and reflective patterns. When joints are examined dynamically the evaluation of respective structures is of great importance.

Sound extinction: Sound extinction corresponds to the interruption of the echogenic cortex layer structure. Characteristically, this is caused by a sound impermeable structure located between the transducer and the interruption of the cortex layer (e.g., metal, wood, bone or calcifications).

Contours: Contours are described as sharp, soft, interrupted, smoothly outlined, infiltrating, not infiltrating.

If the finding is non-pathological, two standard sectional planes of the examined joint must be documented by images and evaluated. When evaluating a pathological finding, it must be imaged in two planes.
- Non-pathological finding: Two standard sectional planes of the organ under examination
- Pathological finding: Presentation of finding by two standard sectional planes possibly with bilateral comparison.

If the finding is ambiguous, an analysis of the contra-lateral view is recommended for comparison. Accurate measurement of size is possible by measuring with calipers. The size of the image on the monitor cannot be related to real size because varying zoom levels may have been applied. Documentation is possible by means of a thermal printer, X-ray film, video, computer (interactive text programs), optical discs (WORM), or others.

7.11 Typical Ultrasound Presentation of Selected Tissue Structures

Bone Surface: Healthy bone surfaces appear as very strong, echogenic reflections (Fig. 7.9).

With the transducer correctly positioned, all standard sectional planes will show the cortical surface as the strongest echo rich reflection on the monitor. The bone surface thus serves as a quality control. Interruptions in the contour of the bone may be caused by an open epiphyseal seam or fracture during the growth phase, or a fracture or lesion following the growth phase. An apparent interruption of the cortex may be observed at any age and may be caused by pseudo-lesions. Changes above the level of the bone cortex may be caused by, e.g., osteophytes or an exostosis. Tumors may also cause osseous changes.

Hyaline Joint Cartilage: Healthy hyaline joint cartilage appears as echo poor or echo free (Fig. 7.10). Distinguishing between hyaline joint cartilage and fluid (synovia or hydrarthrosis), however, is rarely possible because of a boundary sectional plane phenomenon, called "cartilage sign".

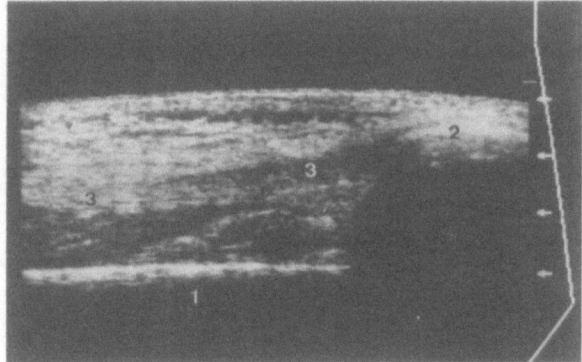

Fig. 7.9. Knee joint, ventral region, supra-patellar longitudinal plane:
1 Femoral cortex,
2 Patellar cortex,
3 Quadriceps tendon

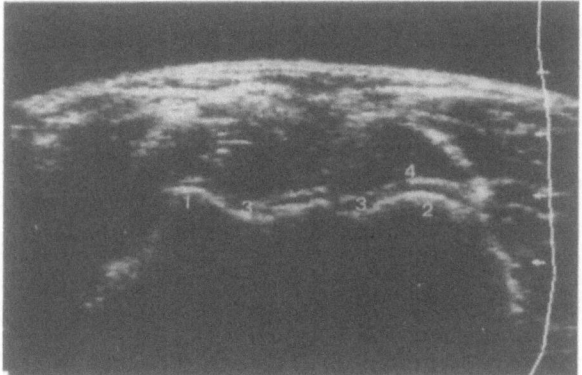

Fig. 7.10. Elbow joint, ventral region – transverse sectional plane:
1 Trocheal cortex,
2 Humeral capitulum cortex,
3 Cartilage,
4 Joint capsule

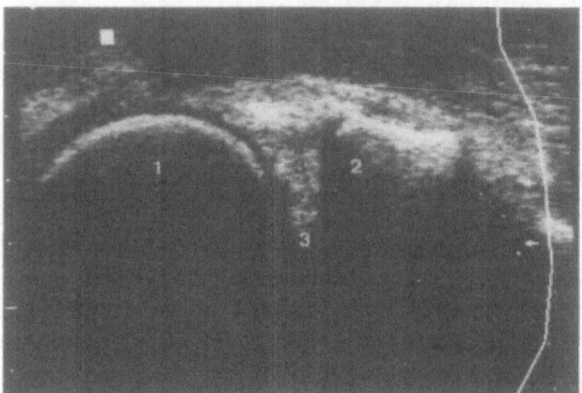

Fig. 7.11. Knee joint, dorsal region – medial longitudinal sectional plane:
1 Medial femoral condyle, 2 Tibia head, 3 Medial meniscus – posterior horn

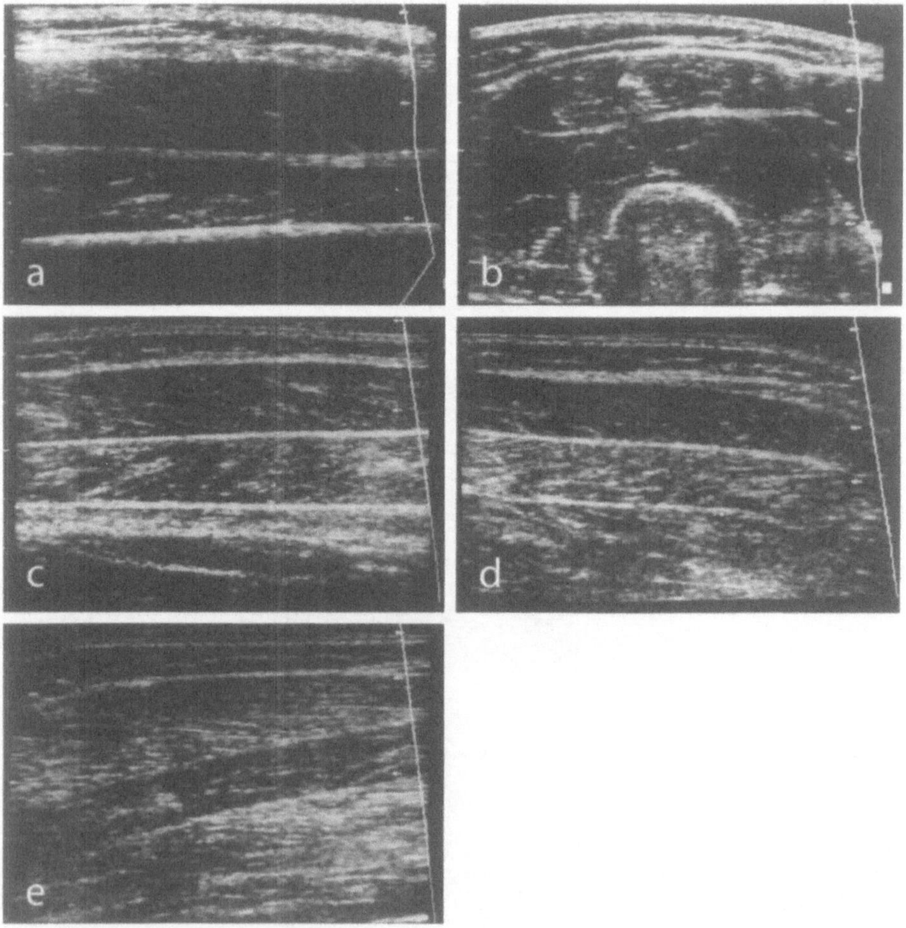

Fibrous Cartilage: Fibrous cartilage presents itself moderately echo rich and homogeneous. Because of the close proximity of the hyaline cartilage and fibrous cartilage, these two structures are relatively easily distinguishable especially in the area of the posterior horn of the meniscus (Fig. 7.11).

Skeletal Muscles: Healthy skeletal muscles produce characteristic images in the longitudinal and transverse sectional plane: The transverse sectional plane shows typical "sprinkles", i.e., areas that alternate between echo rich and echo poor; the longitudinal plane presents typical "feathering" (Fig. 7.12 a–e).

Intramuscular septa appear as either echo rich or echo poor (c–e), depending on the sound impact angle and/or the device setting. In active younger patients, healthy muscles are likely to be echo poor. In the older patient, they are echo rich.

Tendon Tissue: Healthy tendon, when viewed vertically, is echo rich, but echo poor, when the sound direction is inclined (Fig. 7.13 a and b).

When the sound beam direction changes due to the curvature of the tendon, the image can alternate between echo poor and echo rich (Fig. 7.13 c). This is termed reversed reflection or the phenomenon of the wandering reflex or anisotropy. In the interest of reliable ultrasound diagnosis, every tendon should be imaged echo rich to allow clear distinctions between small accumulations of fluid in tenosynovitis and the tendon tissue, for example. Osseous insertions of the healthy tendon appear as echo poor triangular formations. This can be explained by the fact that the tendon fibers bend in different directions, when inserting into the bone. This leads to echo poor regions due to the oblique incidence of sound waves. This can be clearly demonstrated at the insertion point of the Achilles tendon at the calcaneus (Fig. 7.13 d) or in the region of insertion of the M. triceps at the olecranon.

Fatty Tissue: Ultrasound images of healthy fatty tissue appear as homogenous, echo rich structures (Fig. 7.14). Reliable differentiation between healthy fatty tissue and lipo-sarcoma is not possible with ultrasound.

Blood Vessels: Vascular lumen appears practically echo free (Fig. 7.15). The vascular walls, hit vertically by the sound waves, appear echo rich (Fig. 7.15).

Fig 7.12 a–e. Thigh muscles, ventral longitudinal (**a**) and transverse plane (**b**). The healthy skeletal muscles present themselves in a characteristic pattern in the longitudinal as well as in the transverse plane. Both aspects of the M. gastrocnemius show nearly the same echogenity and the same gray value distribution, when the transducer is placed horizontally on the calf (**c**). By changing the position of the transducer (increased pressure on the soft tissue in the proximity of the transducer), the part of the muscle with the vertical incidence of the sound waves will show up echogenic, the other part will appear as echo poor (**d**). If the pressure on the transducer is exerted in the opposite direction, this will also reverse the echogenity distribution of the muscle (**e**). **c–e** Longitudinal sectional plane in the mid-third of the dorsal calf above the M. gastrocnemius caput laterale

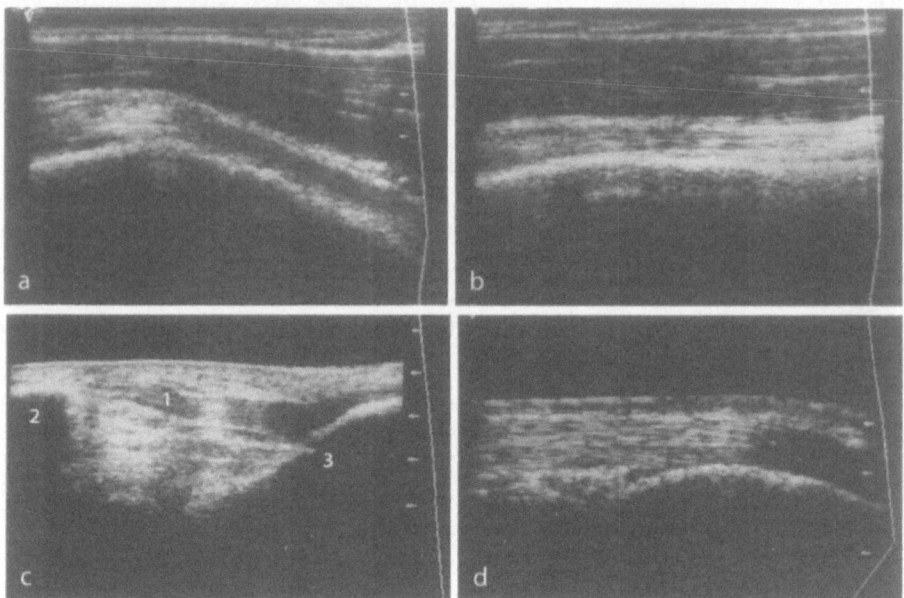

Fig. 7.13 a, b. Shoulder joint, ventral longitudinal sectional plane. The healthy, long head biceps tendon is imaged echo poor when imaged with oblique incidence of the sound beam (**a**), but echogenic when imaged with normal incidence (**b**)

Fig. 7.13 c. Knee joint, ventral infra patellar longitudinal sectional plane. The patellar ligament (1) is imaged 4 months after a rupture and surgery as a continuous structure between the patella (2) and the tibial tuberosity (3) (after Mc Laughlin). Because of its undulating course the patella tendon appears partially either echogenic or echo poor (**c**). Physiologically the insertion area of the large tendon is echo poor and has the shape of a triangle

Fig. 7.13 d. Shows the attachment of the Achilles tendon at the calcaneus in the dorsal longitudinal sectional plane

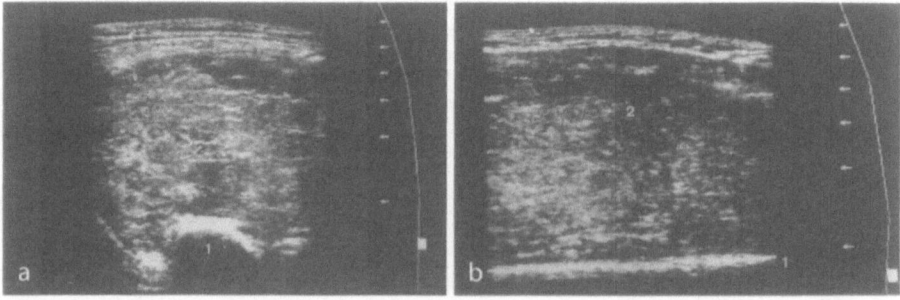

Fig. 7.14 a, b. Distal thigh, ventral region, transverse (**a**) and longitudinal sectional plane (**b**). 1 Femoral cortex layer, 2 Lipoma

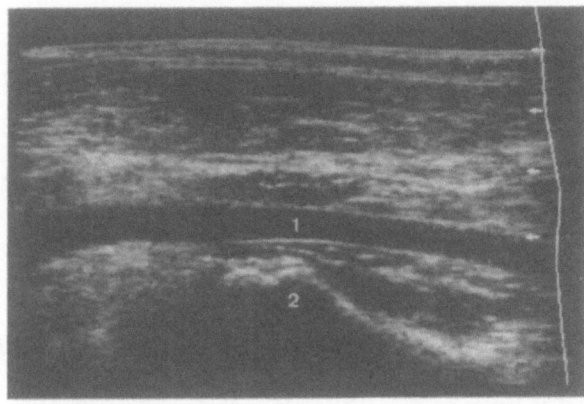

Fig. 7.15. Knee joint, dorsal region – intercondylar longitudinal sectional plane. 1 Femoral artery, 2 Head of tibia

Arteries are clearly recognizable by their characteristic pulsation. For a correct basic setting of the ultrasound device a vessel should be used for calibration of the device: The vascular lumen should show up echo free.

8 Shoulder

8.1 Anatomy and Pathology of the Periarticular Structures of the Shoulder

The shoulder region is composed of the following five functional units:
- Glenohumeral joint
- Acromioclavicular joint
- Sternoclavicular joint
- Scapulo-thoracic space
- Subacromial space.

The glenohumeral joint distinguishes itself by the following characteristics:
- Flat glenoid fossa.
- Relatively large humeral head in relationship to the glenoid fossa.
- Wide range of motion.
- Great importance of soft tissue guidance (rotator cuff) for joint function with only minimal osseous guidance.

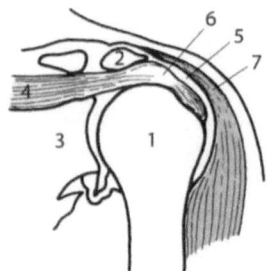

Fig. 8.1. Anatomy of the shoulder
1. Humeral head
2. Acromion
3. Scapula
4. M. supraspinatus
5. Subacromial bursa
6. M. supraspinatus tendon
7. M. deltoid

Pathological changes of the rotator cuff are the most frequent cause of shoulder joint complaints.

The supraspinatus is most likely to exhibit pathology. It is located below, but parallel to the coraco-acromial ligament and the greater tuberosity, and controls the abduction of the humerus. The supraspinatus tendon runs through the isthmus of the subacromial space.

The other structures are the external rotators M. infraspinatus and M. teres minor as well as the M. subscapularis, which is responsible for the internal rotation.

8.2 Standard Ultrasound Examination of the Shoulder

8.2.1 Typical Indications and Findings

Ultrasound examination of the shoulder is most effective for periarticular and intraarticular soft tissue changes. These may for example be inflammatory changes, a new or old tendon lesion in the rotator cuff or a new or old traumatic bone lesion (e.g., greater tuberosity deformity, Hill-Sachs lesion). Table 8.1 lists the common and rare indications for ultrasound examination of the shoulder joint.

8.2.2 Examination

The patient should be positioned sitting upright in a chair without an arm rest, facing the monitor between the examining physician and the ultrasound device. The crucial advantage of shoulder joint sonography as compared to the examination of other joints is the outstanding possibility to do dynamic imaging. Often only in this way is a reliable diagnosis possible. First, the device is set for the appropriate standard sectional plane. The shoulder joint is then dynamically examined by rotating the upper arm, remaining in the same sectional plane. Examination of the shoulder joint may be performed

Table 8.1. Typical indications and findings

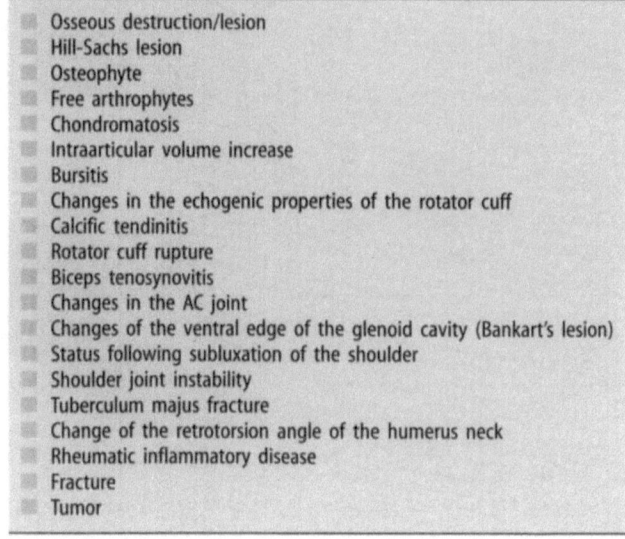

- Osseous destruction/lesion
- Hill-Sachs lesion
- Osteophyte
- Free arthrophytes
- Chondromatosis
- Intraarticular volume increase
- Bursitis
- Changes in the echogenic properties of the rotator cuff
- Calcific tendinitis
- Rotator cuff rupture
- Biceps tenosynovitis
- Changes in the AC joint
- Changes of the ventral edge of the glenoid cavity (Bankart's lesion)
- Status following subluxation of the shoulder
- Shoulder joint instability
- Tuberculum majus fracture
- Change of the retrotorsion angle of the humerus neck
- Rheumatic inflammatory disease
- Fracture
- Tumor

with the arm hanging free or with the elbow joint flexed at a right angle. In special situations, such as for Bankart's lesion or for retro-torsional angle analysis of the humerus, a functional examination is required. If the shoulder joint movement is restricted, moving the transducer may permit an adequate evaluation of the joint.

8.2.3 Standard Sectional Planes

Three regions of the shoulder joint can be examined sonographically using a total of six standard sectional planes. In each region, the joint is imaged in two sectional planes, which are almost vertical to each other:

- Dorsal region
  - Transverse sectional plane
  - Longitudinal sectional plane
- Lateral-superior region
  - Transverse sectional plane
  - Longitudinal sectional plane
- Ventral region
  - Transverse sectional plane
  - Longitudinal sectional plane.

If the finding is non-pathological, it is recommended that the two standard sectional planes of the lateral superior region be documented. Shown below are transducer position, normal ultrasound image and a explanatory diagram for every standard sectional plane.

1) Dorsal Region – Transverse Sectional Plane

Transducer position: The transducer is in the region of the infraspinatus fossa, parallel to the scapular spine. The transducer is ascending from medial caudal to lateral cranial with its lateral edge placed above the dorsal region of the humeral head.

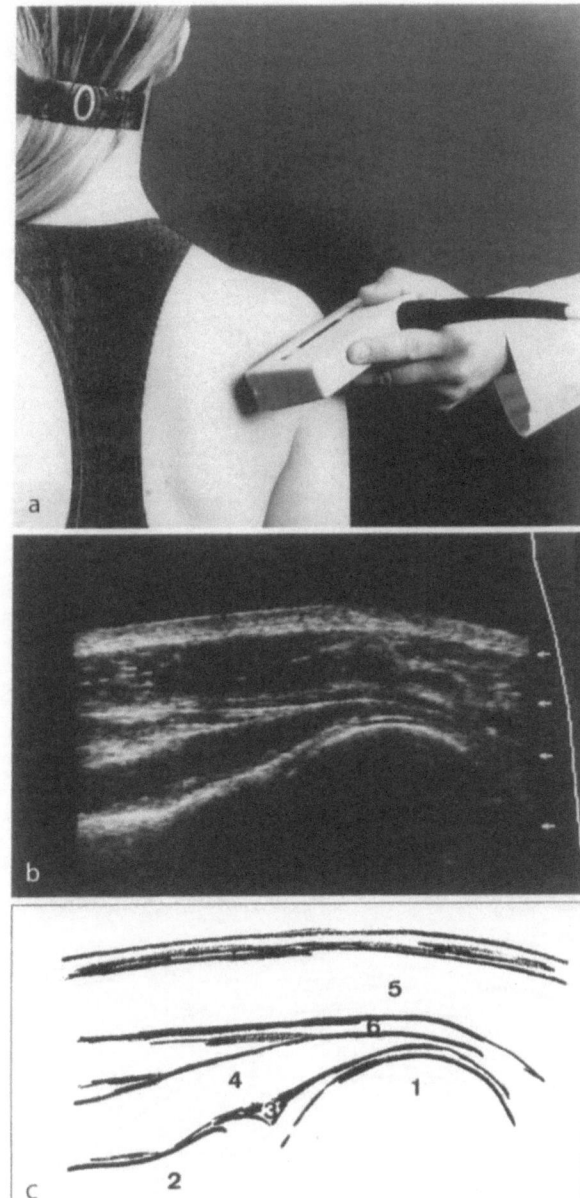

Figure 8.2 a–c. Shoulder joint, dorsal region – transverse sectional plane.
b, c.
1 Humeral head,
2 Scapula,
3 Glenoid labrum posterior,
4 M. infraspinatus,
5 M. deltoid,
6 Subdeltoid bursa

2) Dorsal Region – Longitudinal Sectional Plane

Transducer position: The transducer is positioned parallel to the axis of humeral shaft. The transducer is placed superior to the dorsal region of the rotator cuff with the cranial region resting on the acromion. The upper arm should be maximally internally rotated resting against the body.

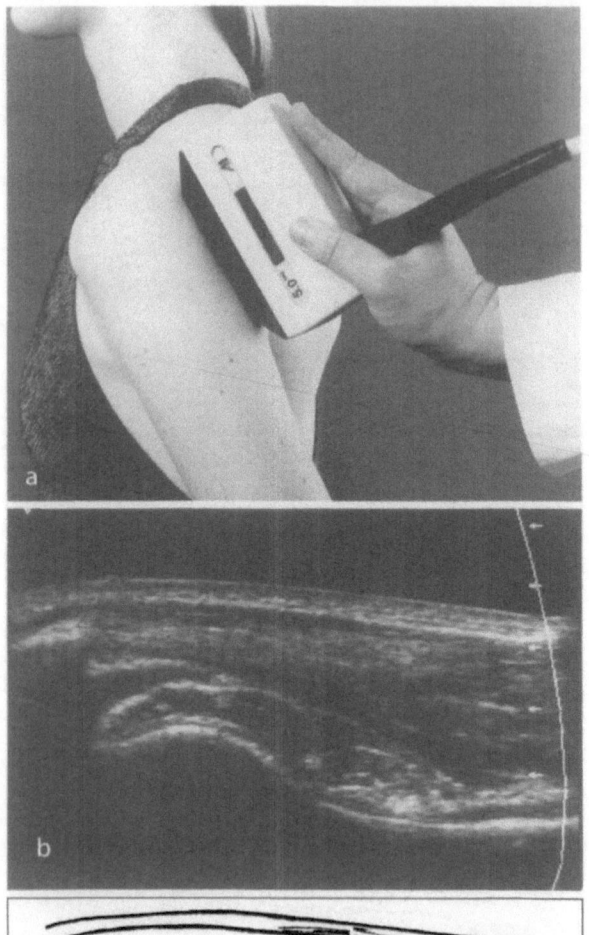

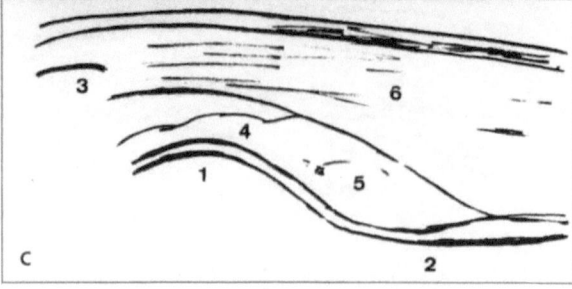

Fig. 8.3 a–c. Shoulder joint, dorsal region – longitudinal sectional plane.
b, c.
1 Humeral head,
2 Humeral shaft,
3 Acromion,
4 M. infraspinatus,
5 M. teres minor,
6 M. deltoid

3) Lateral Superior Region – Transverse Sectional Plane

Transducer position: The transducer is placed lateral-caudal and parallel to the coracoacromial ligament.

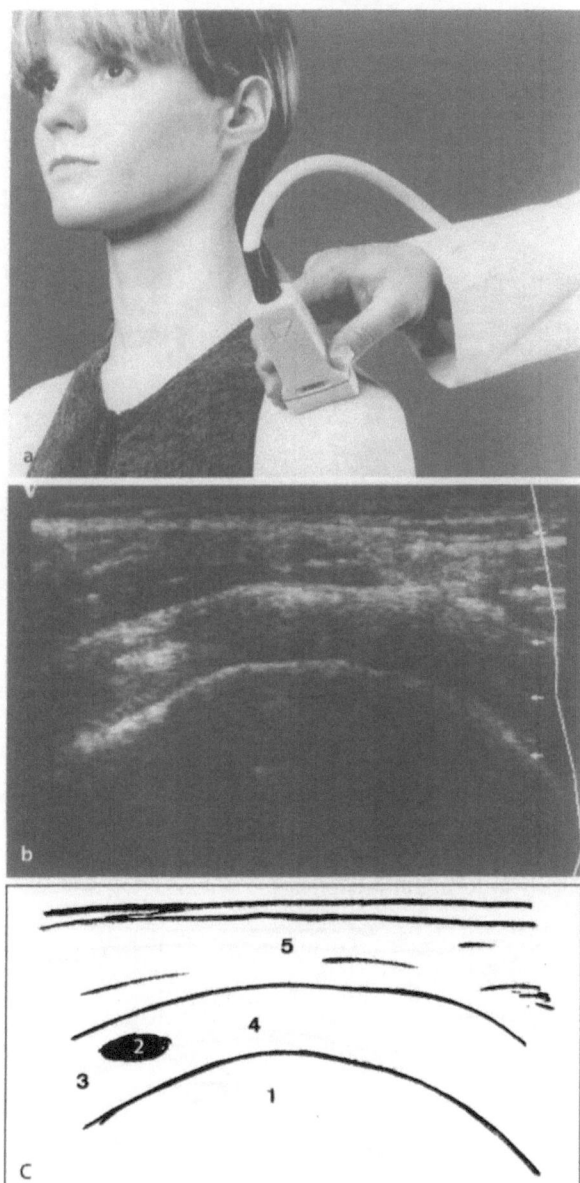

Fig. 8.4 a–c. Shoulder joint, lateral superior region – transverse sectional plane.
b, c.
1 Humeral head,
2 Long biceps tendon,
3 M. subscapularis,
4 M. supraspinatus,
5 M. deltoid

4) Lateral Superior Region – Longitudinal Plane

Transducer position: The transducer is placed in line with the trapezoid muscle and superior to the acromion and the lateral region of the supraspinatus tendon.

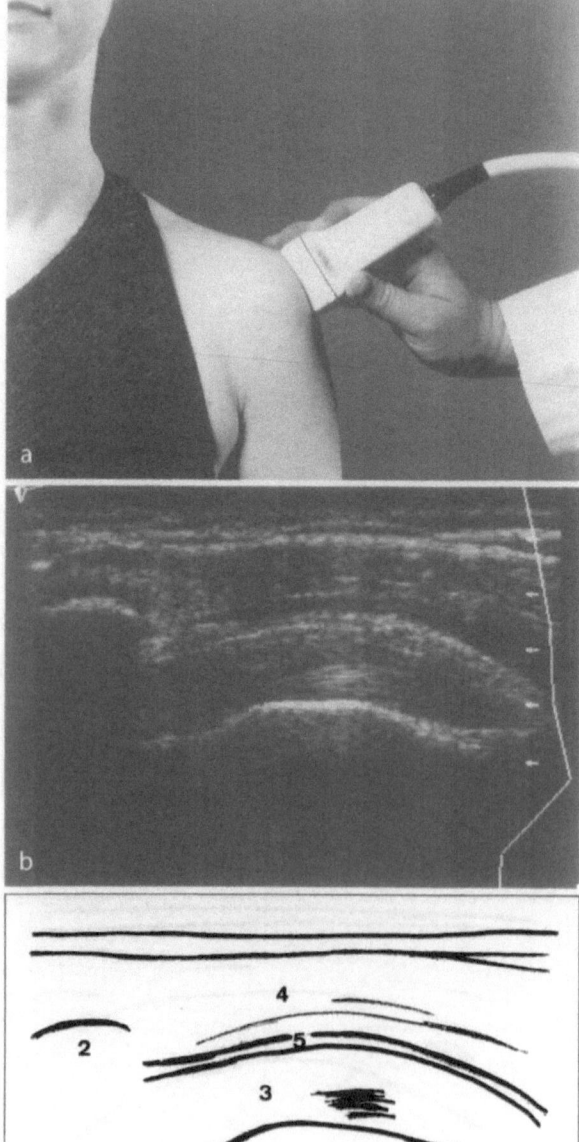

Fig. 8.5 a–c. Shoulder joint, lateral superior region – longitudinal sectional plane.
b, c.
1 Humeral head,
2 Acromion,
3 M. supraspinatus,
4 M. deltoid,
5 Subdeltoid bursa

5) Ventral Region – Transverse Sectional Plane

Transducer position: The transducer is transverse above the ventral regions of the proximal humerus. Transducer is placed from ventral, strictly horizontal, immediately superior to the sulcus intertubercularis.

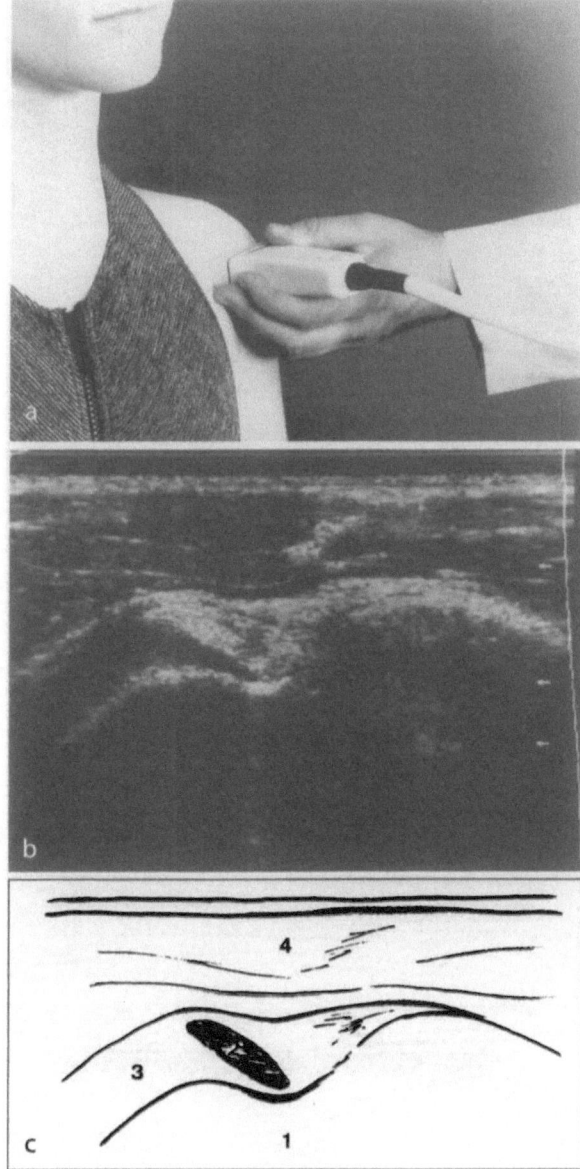

Fig. 8.6 a–c. Shoulder joint, ventral region – transverse sectional plane.
b, c.
1 Humeral head,
2 Long biceps tendon,
3 M. subscapularis,
4 M. deltoid

6) Ventral Region – Longitudinal Sectional Plane

Transducer position: Upon completion of the ventral transverse sectional plane, the upper arm should be positioned so that the long biceps tendon appears at the top of the monitor screen. This facilitates the adjustment of the ventral longitudinal sectional plane. The transducer is parallel to the longitu-

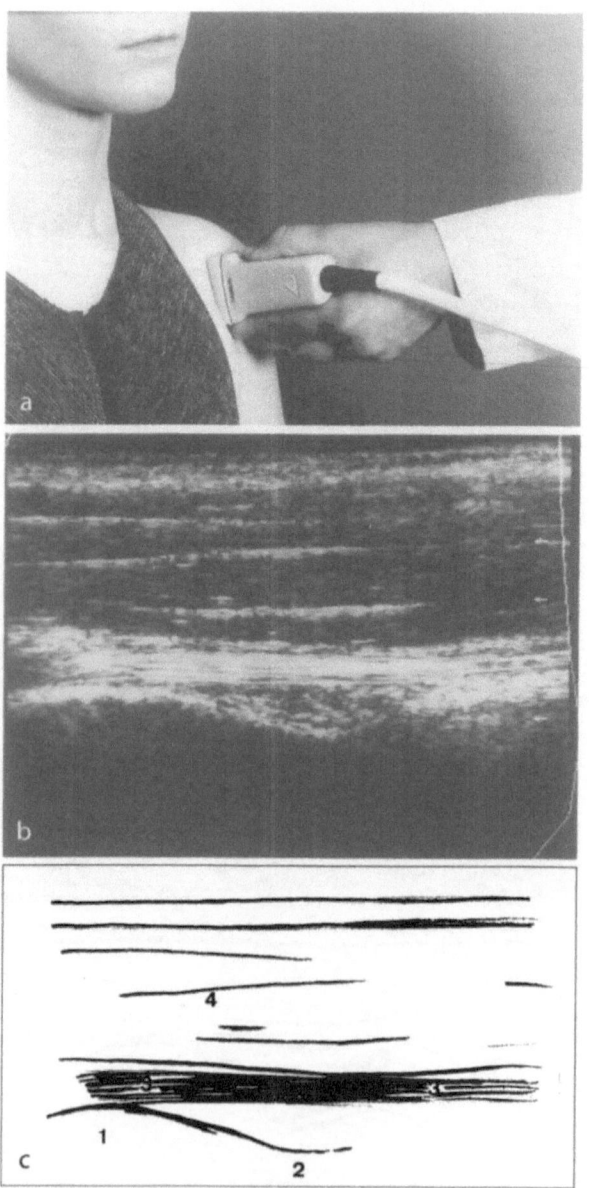

Fig. 8.7 a–c. Shoulder joint, ventral region – longitudinal sectional plane.
b, c.
1 Humeral head,
2 Humeral shaft,
3 Long biceps tendon,
4 M. deltoid

dinal axis of humeral shaft. The transducer is placed superior to the ventral sectional planes of the humerus.

8.3 Calcific Tendinitis

8.3.1 Definition

Calcific tendinitis is defined as a hydroxylapatit deposit near the insertion area of the rotator cuff.

8.3.2 Pathology

The calcific tendinitis of the shoulder joint is an independent disorder. Etio-pathogenetically, two concepts are discussed: In the past, calcified necrosis due to mechanical, vascular, biochemical or genetic factors was favored; to-day however, the concept of calcification following metaplasia with active cal-cification in the vital tendon is more widely accepted.

The disorder follows a four phase cycle: In the *transformation phase*, me-taplasia from tenocytes to chondrocytes is observed. In the *calcification or formation phase*, hydoxylapatit crystals are imbedded in the intercellular sub-stance of the cartilage cell. In the *absorption phase*, an ingrowth of vessels due to an unknown stimulus followed by the phagocitosis of the crystals is observed. In the *repair phase*, new tendon tissue is formed. Normally, the calcium deposit can be found in the hypo-vascular zone between the fibrous

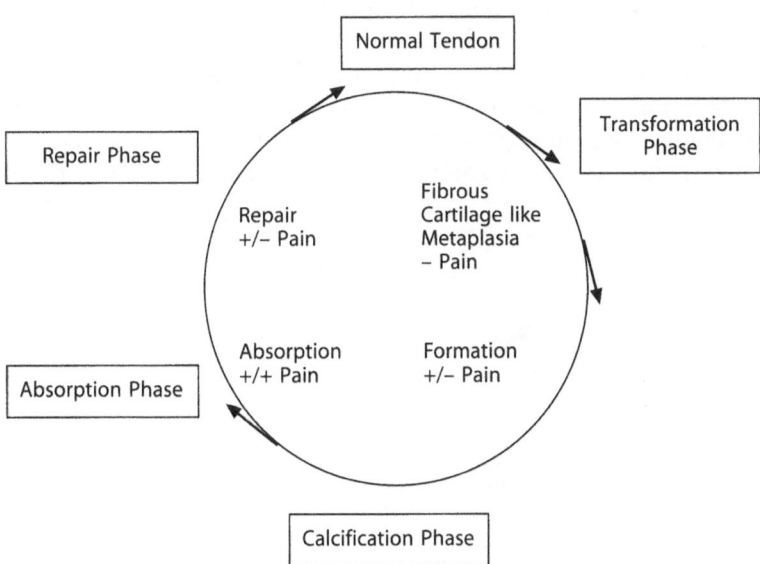

Fig. 8.8. Etiologic model [Uhthoff and Sarkar (1981)]

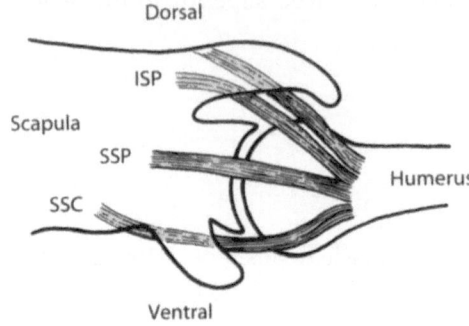

Dorsal

ISP

Scapula

SSP

SSC

Humerus

Ventral

Fig. 8.9. View of the shoulder joint from above; scapula (shoulder blade), humerus, ventral (front)

cartilage region and the tendon insertion point of the rotator cuff. A spontaneous absorption takes place in 9.3–27.1% of all cases.

The supraspinatus tendon, in the area of the insertion point of the greater tuberosity, is most often (approx. 90%) affected. But the calcium deposit may also be observed in any other tendon or in several tendons at the same time, even in the long head biceps tendon.

The middle-aged population (age 40–50) is most often affected, affecting more females than men (1.2–2:1), and primarily the right shoulder joint (1.2–1.7:1). Bilateral calcifications account for 8.8–40% of all cases.

8.3.3 Clinical Diagnosis

The calcific tendinitis can develop either asymptomatically or highly acute. With regard to the clinical symptomatics two phases are differentiated: acute and chronic.

During the chronic phase changing pain severity is observed, from no pain to very painful. Often at night when lying on the shoulder, pain is reported in the affected shoulder as well as in the upper arm, accompanied by a restriction in function and a feeling of weakness in the shoulder joint. During clinical examinations, a painful projection to the distal region of the dermatom C5, which corresponds to the insertion point of the M. deltoid, is found.

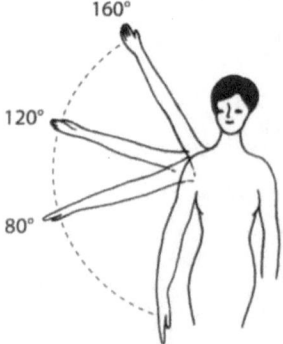

160°

120°

80°

Fig. 8.10. Painful arc

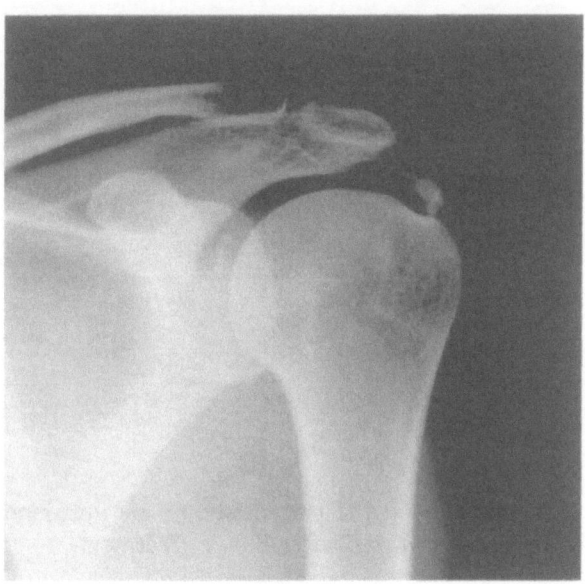

Fig. 8.11. Calcium deposits typically form in the region of the greater tuberosity; X-ray image

Most frequently there is a painful arc between about 60 and 120 degrees of abduction. The movement in the glenohumeral joint is either free or restricted according to a cuff pattern (abduction>external rotation>internal rotation). Verbal response to pain upon palpation is often non-specific and thus not helpful. Impingement tests (according to Neer or Hawkins and Kennedy or Matsen, etc.) are usually positive. Isometric resistance tests often show a painful reaction: Abduction tests the M. supraspinatus, external rotation the M. infraspinatus and internal rotation the M. subsapularis.

During the acute phase, pain may start instantaneously, often described as a stabbing, pounding or throbbing pain. At clinical examination the signs of the chronic phase are usually more pronounced. The arm often cannot be moved due to pain and must be supported by the opposite hand.

During the absorption phase, disappearance of the calcium deposit is observed either instantaneously or immediately following shock wave therapy. In this phase, highly acute syndromes are observed for several weeks. Following the absorption of the calcium deposit, residual problems may persist for several months, e.g., in the case of overhead activities.

In this context it is important to inform the patient that the absorption phase with highly acute complaints may be initiated by shock wave treatment and that initiation of the absorption phase is an objective of shock wave therapy.

If the calcium deposit penetrates the bursa, which can occur spontaneously or following shock wave therapy, the patient will suffer intensified pain *for a short period*. In some cases, the faster absorption of the calcium deposit in the bursa is also initiated by shock wave therapy.

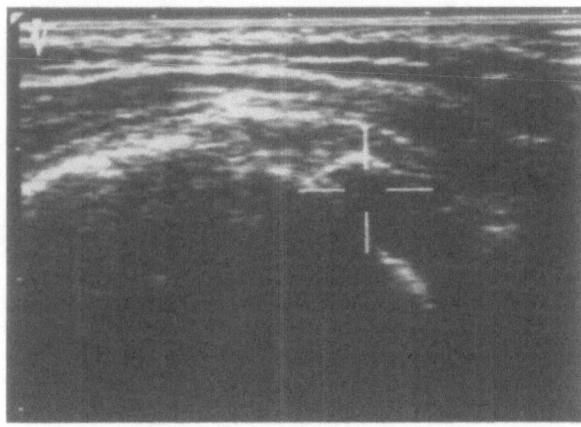

Fig. 8.12. Calcium deposits with typical localization in the region of the greater tuberosity; ultrasound image

In this context the patient should be informed of the possibility of short-term intensified pain after ESWT treatment.

8.3.4 X-ray

According to the classification by Gaertner, three types are radiologically differentiated: For type 1 the calcium deposit is dense with well-defined boundaries. This type is found in the calcification phase. Type 2 is a "mixed type"; the calcium deposit is either dense and diffuse at the boundary or transparent and sharp bordered. Type 3 is observed in the absorption phase, the calcium deposit is transparent and diffuse at the boundaries.

8.3.5 Ultrasound

Calcifications do not show a unique image pattern. Depending on the consistency, the image either displays echogenic reflexes with or without sound weakening/shadow or inhomogeneities of the tendon without sound weakening.

The macroscopically hard calcific deposits, which appear on the X-ray image with a well-defined boundary and dense shadows (according to X-ray classification: type 1 Gaertner), sonographically present a clear sound shadow, in addition to a strong echogenic reflex.

Macroscopically fluid, pasty calcifications, which have diffuse boundaries and are transparent on the X-ray image, do not reflect sound waves totally. They are imaged either as echogenic reflexes or as inhomogeneities of the tendon; the guiding sound weakening is missing (according to X-ray classification: type 2 or 3 Gaertner).

Sonographically, during the dynamic examination of the calcium deposit, movement is performed in the direction of the respective tendon of the rotator cuff. Thus, a distinct differentiation of a calcification in the bursa and of an osseous tear of the rotator cuff is possible. By rotating the upper arm, the calcium deposit can be positioned in the corresponding plane.

Multiple calcifications can be identified in different standard sectional planes. In the lateral superior longitudinal sectional plane, the calcification of the infraspinatus is generally positioned laterally, while the calcifications of the supraspinatus tendon are located more centrally. Especially the transverse sectional planes simplify the topographical classification of calcifications with respect to the individual tendons of the rotator cuff.

8.3.6 Conservative and Surgical Therapy

Therapeutic recommendations comprise nearly all conservative and surgical orthopedic treatment options.

Conservative Treatment:
- Physiotherapy, proprioceptive training
- Systematic drug therapy, application of ointments
- Subacromial injection with local anesthetic, possibly with addition of steroids
- Physical therapy, electro-therapy, thermal therapy, massage
- X-ray stimulation treatment
- Needling.

Surgical Treatment:
Surgery is recommended when the pain continues to persist following conservative treatment and shock wave therapy. Various surgical techniques can be used with the goal of removing the calcium deposit through miniarthrotomy or arthroscopy. Additionally, acromio plasty can be performed simultaneously, if the subacromial space is narrowed.

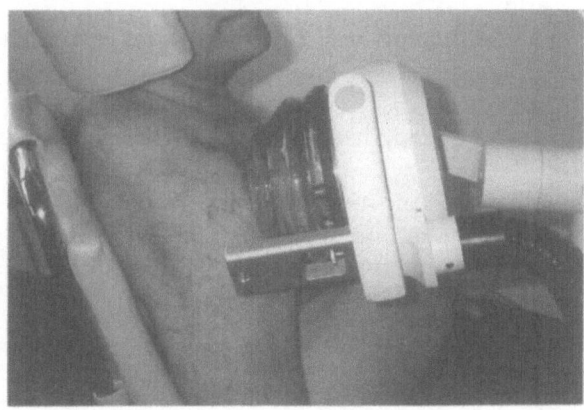

Fig. 8.13. ESWT of calcific tendinitis, localization with patient feedback

8.3.7 ESWT

1) Prerequisites

1. The patient should have undergone a minimum of six months of conservative therapy.
2. Patient information given and written patient consent form (see above) signed.
3. Exclusion of contraindications (see chapter 4.4).
4. ESWT should be followed by proprioceptive training and physiotherapy and should focus on strengthening the internal and external rotators and adductors to optimize the positioning and stabilization of the humeral head.

2) Anesthesia

Contrary to high energy ESWT, anesthesia is usually not required for low energy ESWT. If required, a subacromial injection or plexus anesthesia should be considered.

3) Positioning and Localization

Positioning is dependent on the region of treatment and the localization system used. The patient can be in a sitting, prone, or supine position.

Patient Feedback (Laser Pointer)

Positioning:
 The patient is sitting in a chair (without arm rests).
   – The therapy head is coupled from ventral or from dorsal with the arm hanging free and with slight internal or external rotation.

Remember the coupling gel!

 The patient is in supine position with internal or external rotation of the arm.
 – The therapy head is coupled from ventral and superior to the patient.
 The patient is in prone position with internal or external rotation of the arm.
   – The therapy head is coupled from dorsal and superior to the patient.

Localization: Palpation and marking of the ROI (X). Determination of the ventral coupling point for the therapy head (O). Bring the laser pointer and point of maximum pain (X) into alignment (= definition of penetration depth).

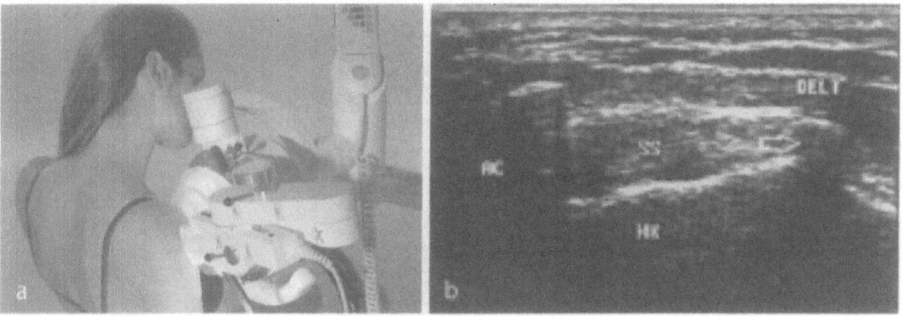

Fig. 8.14 a, b. Coupling and ultrasound image of a calcium deposit in proximity to the insertion point of the supraspinatus tendon, lateral-superior longitudinal sectional plane ("Bird beak sectional plane")

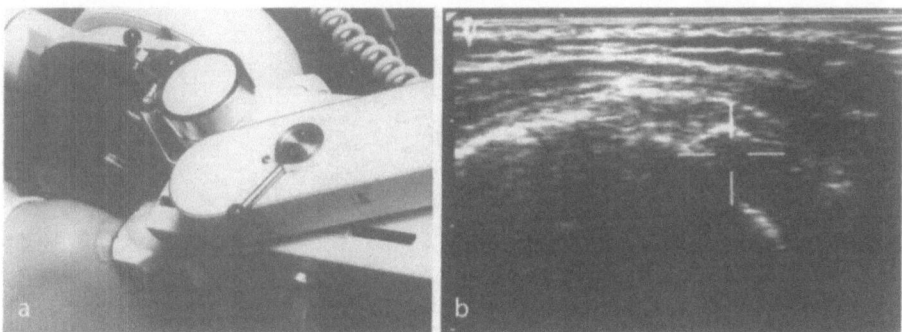

Fig. 8.15 a, b. Ultrasound localization: Lateral superior transverse sectional plane ("Wagon wheel sectional plane")

Ultrasound

Positioning:
▒ The patient is sitting in a chair (without arm rests).
  - The therapy head is coupled from ventral, the arm hanging free and with slight internal or external rotation.
  - Ultrasound coupling in the lateral superior longitudinal sectional plane.

Remember the coupling gel!

▒ The patient is in a supine position on the treatment table with the arm internally or externally rotated.
  - The therapy head is coupled from ventral and superior to the patient.
  - Lateral coupling of the ultrasound transducer.

In some patients, coupling of the therapy head from dorsal may be advantageous.

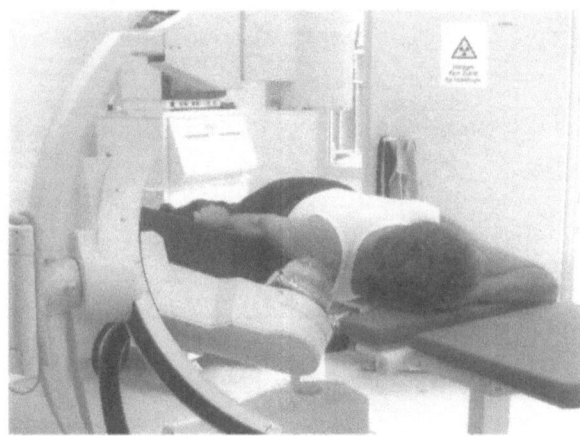

Fig. 8.16. ESWT treatment of calcific tendinitis with X-ray localization

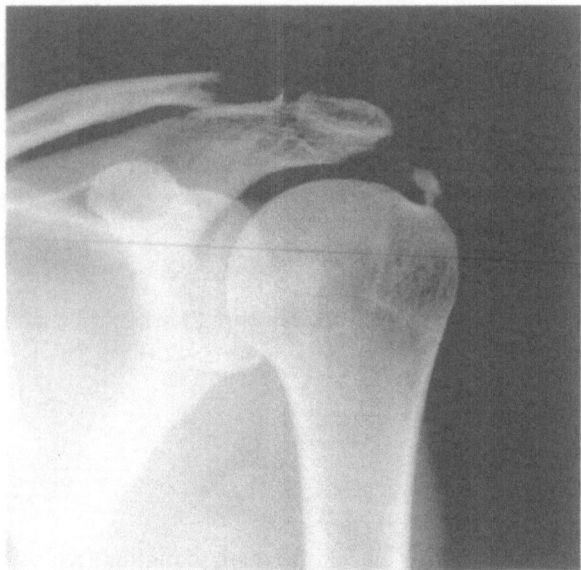

Fig. 8.17. X-ray image of a calcium deposit

Localization:

Lateral-superior longitudinal sectional plane: The sectional plane corresponds to the lateral extension of the M. trapezius and the acromion, along the M. supraspinatus. This sectional plane is referred to as the "bird beak sectional plane", because the shape of the M. supraspinatus sonographically resembles the beak of a bird.

Lateral-superior transverse sectional plane ("Wagon wheel sectional plane"): By rotating the transducer by 90 degrees, while maintaining the image of the calcium deposit on the monitor, a second sectional plane can be established and, if necessary, readjusted.

X-ray Localization

Positioning: The use of X-rays requires the patient to be positioned on a treatment table.
- Prone position.
- Treatment side faces the therapy head.
- Arm close to the body and internally or externally rotated.
- Shoulder is positioned in the treatment table cut-out.
- Patient's head is rotated to the contra-lateral side.

Remember the coupling gel!

In the case of X-ray imaging, observe radiation protection for the patient.

Localization: Imaging of the calcium deposit first in anterior/posterior direction. After alignment of the crosshair with the ROI, tilt the C-arm and adjust in the second plane.

4) Coupling

Tangential Coupling Using the Two Point Method (TPM):
1. Position patient (see above).
2. Find and mark the tender point on tendon = penetration depth of the shock wave (X).
3. Project position of calcium deposit to anterior side of shoulder and mark = entrance point of the shock wave (O).
4. Coupling of therapy head to the ventral mark.
   Remember the coupling gel!
5. With isocentric X-ray C-arm or ultrasound transducer (coupling laterally with longitudinal and transverse presentation) precise targeting on the tendon.

Direct Coupling Using the One Point Method (OPM):
1. Position patient as above.
2. Find and mark the tender point = entrance point of shock wave.

CAUTION: Injury to the lungs may occur if penetration depth is too deep!

5) Treatment Regimen

Once exact X-ray or ultrasound localization has been completed, shock wave therapy can begin. The required energy level can be reached by gradually increasing the energy level depending on the patient's sensitivity to pain and the reaction to analgesic effects caused by shock wave application.

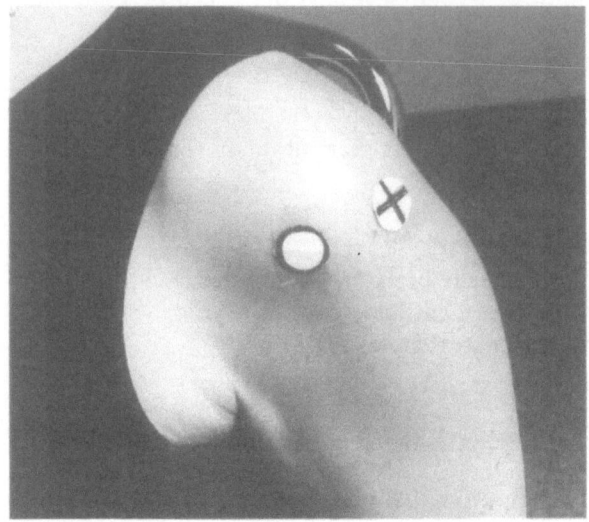

Fig. 8.18. Tangential coupling (TPM)

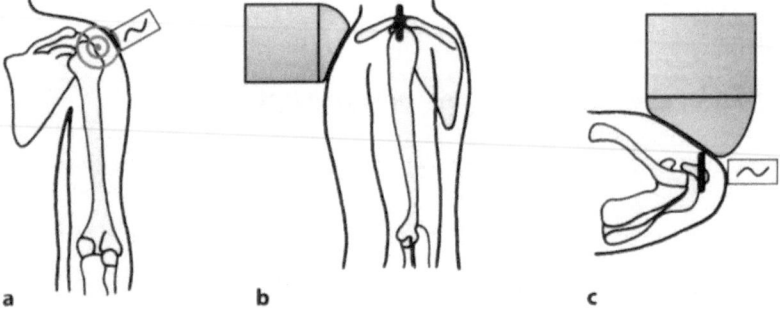

a   b   c

Fig. 8.19. Tangential coupling, presented from different directions. **a** Ventral sectional plane, **b** Lateral sectional plane, **c** From above

The energy level has to be selected according to the specific indication and the objective of the treatment: For pain treatment, only low energy therapy is sufficient. Medium energy application is indicated if a calcium deposit must be fragmented.

X-ray localization should be verified at least twice during shock wave application. With ultrasound continuous, real-time localization control is possible.

The absolute number of pulses and energy level is determined by the therapy concept of the treating physician.

An application of 1500–2000 shock waves per therapy session is recommended.

The recommended cumulated positive energy flux density per treatment summed up over all sessions should be approximately 1300 mJ/mm^2.

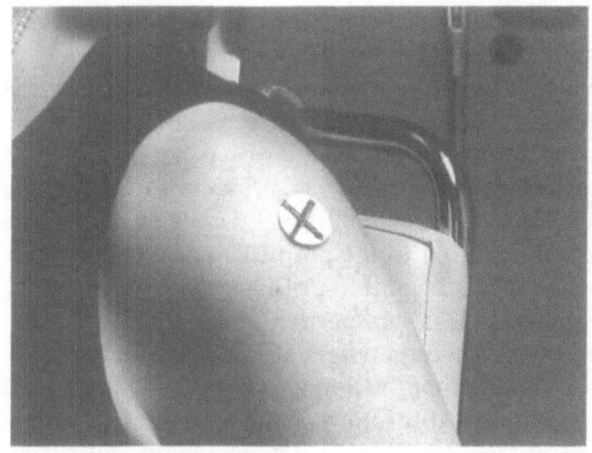

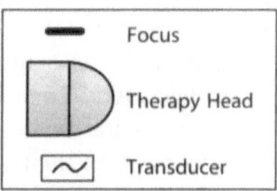

Fig. 8.20. Direct coupling (OPM)

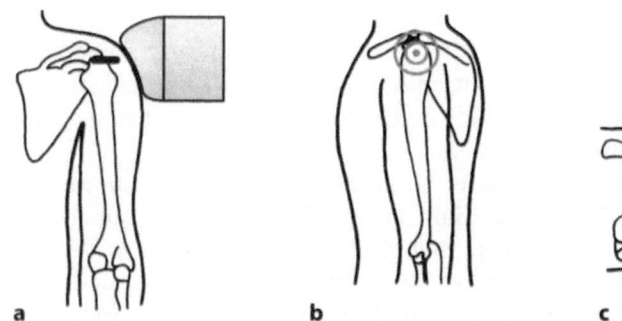

Fig. 8.21. Direct coupling, view from different directions. **a** Ventral sectional plane, **b** Lateral sectional plane, **c** Direct view

8.4 Results

Table 8.1. Recently published results of shock wave therapy of calcific tendinitis

Author	Patients	Follow-up	Success (subjective) %	Success (radiological) %	Energy
Loew 1999	20	3 months	30	20	1×2000, 0.1 mJ/mm^2
Loew 1999	59	6 months	70	77	2×2000, 0.3 mJ/mm^2
Rompe 1998	50	24 weeks	68	64	1×1500, 0.08 mJ/mm^2
Gerdesmeyer 2000	963	17 months	73.6		2.3×2000, 0.28 mJ/mm^2
Rompe 1997	100	24 weeks	67	57	1×1500, 0.28 mJ/mm^2
Daecke 1997	115	6 months	54	77	2×2000, 22 KV
Buch 1996	15	6 weeks	54	66	2×2000, 0.28 mJ/mm^2
Buch 1996	15	6 weeks	20	6.6	5×3000, 0.08 mJ/mm^2
Loew 1999	20	3 months	30	20	1×2000, 0.1 mJ/mm^2

Since the publication of the pilot study by Loew at al., which reports about initial experiences with shock wave therapy for calcific tendonitis, this indication has evolved together with other indications for ESWT such as plantar fasciitis, lateral epicondylitis and pseudarthrosis. Following the first positive results by Loew, numerous studies have been performed. For example, in 1997 Rompe et al. reported that after medium energy ESWT significant improvement or complete freedom of pain was achieved in 68% of the treated patients, applying one treatment of 1500 shock waves at an average flux density of 0.28 mJ/mm^2. In addition to the positive clinical results, they showed complete absorption or at least significant disintegration of the calcific deposit in 64% of the patients. In a prospective, randomised, placebo-controlled study Loew et al. investigated the effect of ESWT on the development of calcific tendonitis a second time, verifying the formerly published observations. In their study, complete absorption of the calcific deposit or a significant disintegration was observed in 77% of patients. Similarly in 70% of the cases complete freedom from symptoms or nearly complete reduction of pain was reported after applying 2000 shock waves each in two sessions at 0.3 mJ/mm^2. In a large population of 963 patients, Gerdesmeyer et al. confirmed this result. Applying medium energy shock waves of 0.28 mJ/mm^2, they achieved complete freedom from symptoms or nearly complete reduction of pain in 73% of cases. In addition, they also achieved stable results over a longer period of time with 2 year follow-up results being identical to short-term results after 6 months.

In comparative studies Loew et al. as well as Buch et al. showed that the application of ESWT using medium energy is significantly superior to low energy ESWT with respect to radiological results as well as with respect to subjective results. In the study by Loew, 20 patients were treated once with 2000 shock waves and 0.1 mJ/mm^2. In only 30% of the cases were good and very good results and in only 20% complete disintegration of the calcific deposit was achieved. A similar result was achieved by Buch in his study comparing two populations treated with medium and low energy ESWT. The superiority of medium energy ESWT was more clear than in the study by Loew et al.

In most publications the authors report about slight side effects in the form of small local petechial skin bleedings and seldom about small hematoma, which heal without complications. As long as general guidelines for the application of ESWT are observed, more serious side effects have not been reported.

9 Elbow

9.1 Standard Ultrasound Examinations of the Elbow

9.1.1 Indications and Findings

Periarticular soft tissue changes and intraarticular changes can be detected by ultrasound. In this context inflammatory changes similar to those seen in rheumatoid patients are of most importance. Table 9.1 lists common and rare indications for ultrasound examination of the elbow.

Table 9.1. Typical indications and clinical findings

Osseous destruction/lesion
Osteophyte
Free arthrophytes
Chondromatosis
Panner's disease
Intraarticular volume increase
Bursitis
Radius head luxation
Elbow joint instability
Changes of humerus neck retro-torsion angle
Inflammatory and rheumatoid diseases
Fracture
Tumor

9.1.2 Examination

The patient should be sitting adjacent to the ultrasound device, facing the monitor screen. To adjust for the ventral sectional plane, the elbow joint should be supine and extended; for the dorsal sectional plane, the elbow joint should be flexed at a right angle. It is helpful if the patient's internally rotated upper arm rests on his/her thigh of the same side. In the presence of a shoulder joint disorder, the examination can be performed with the patient in the supine position: For ventral sectional plane adjustment, the arm should lie supine on the treatment table, next to the upper part of the body.

For the dorsal sectional plane adjustment, the arm with the elbow joint flexed at a right angle should be placed on the patient's abdomen.

9.1.3 Standard Sectional Planes

Ultrasound examination of the elbow is performed in two regions and in five standard cross-sectional planes. In each region, the joint is imaged in two or three almost vertical sectional planes:

- Ventral region
  - Transverse sectional plane
  - Humero radial longitudinal sectional plane
  - Humero ulnar longitudinal sectional plane.
- Dorsal region
  - Transverse sectional plane
  - Longitudinal sectional plane.

If the finding is not pathological, documentation of the following two sectional planes is recommended:

- Ventral transverse sectional plane
- Dorsal longitudinal sectional plane.

In the following for each standard sectional plane, the transducer position, a standard ultrasound image and an explanatory diagram are presented.

1) Ventral Region – Transverse Sectional Plane

Transducer position: The transducer is positioned at the distal humerus in the transverse sectional plane. The transducer is guided from the upper distal third of the humerus toward the elbow joint and is positioned superior to the trochlea and the humeral capitulum.

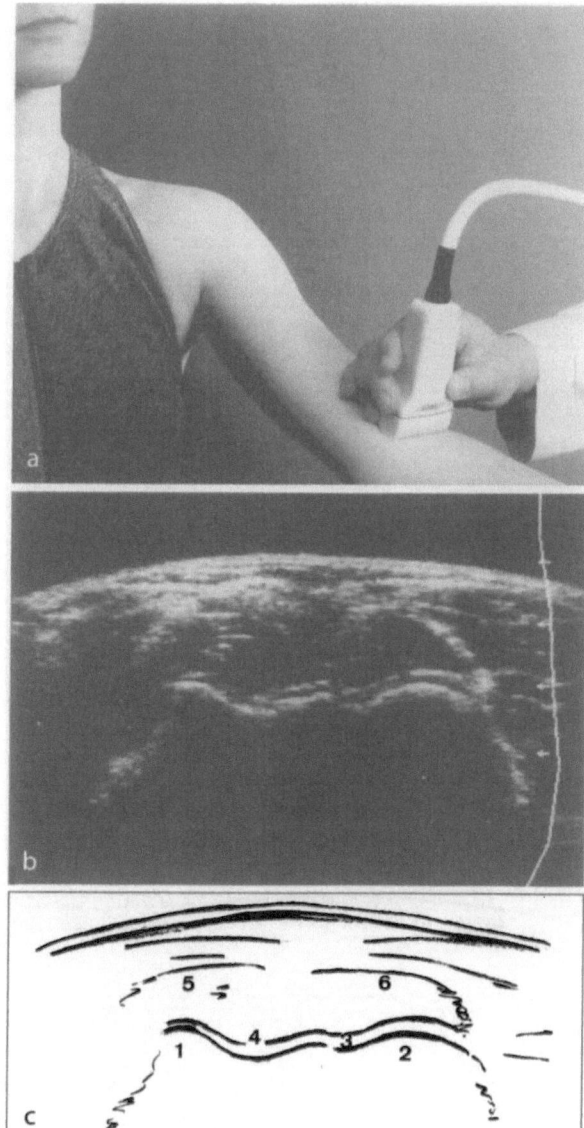

Fig. 9.1 a–c. Elbow joint, ventral region – transverse sectional plane.
b, c.
1 Humeral trochlea,
2 Humeral capitulum,
3 Hyaline joint cartilage,
4 Joint capsule,
5 M. brachialis,
6 M. brachioradialis

2) Ventral Region – Humero Radial Longitudinal Plane

Transducer position: The transducer position is on the side of the radius superior to the elbow in the longitudinal sectional plane. The transducer is placed parallel to the humeral and radial shaft axis. These osseous guiding structures are positioned parallel to the upper monitor edge.

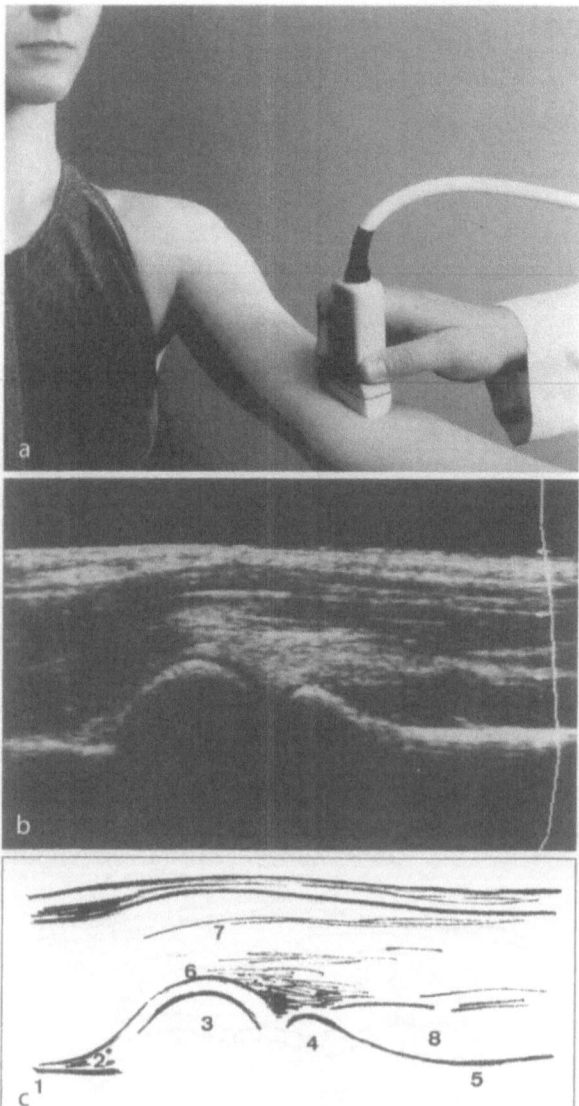

Fig. 9.2 a–c. Elbow joint, ventral region – humero radial longitudinal sectional plane.
b, c.
1 Humeral shaft,
2 Radial fossa,
3 Humeral capitulum,
4 Radial head,
5 Shaft of radius,
6 Joint capsule,
7 M. brachioradialis,
8 M. supinator

3) Ventral Region – Humero Ulnar Longitudinal Plane

Transducer position: The transducer is positioned on the ulnar side superior to the elbow joint in the longitudinal sectional plane and parallel to the humeral and the ulnar shaft axis. These osseous structures are positioned parallel to the upper edge of the monitor.

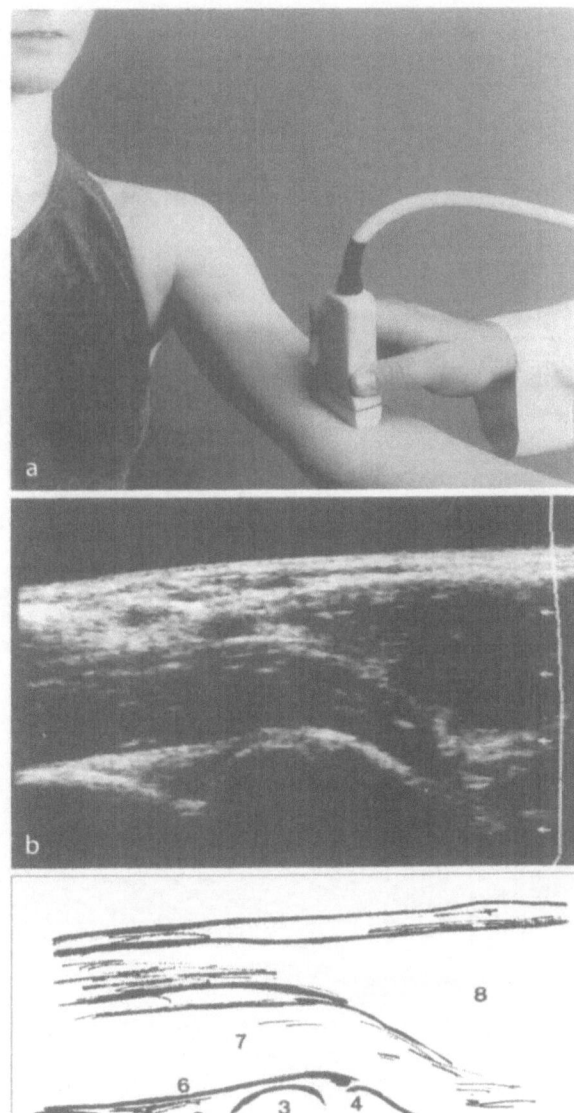

Fig. 9.3 a–c. Elbow joint – humero ulnar longitudinal sectional plane.
b, c.
1 Humeral shaft,
2 Coronoid fossa,
3 Humeral trochlea,
4 Coronoid process,
5 Ulnar shaft,
6 Joint capsule,
7 M. brachialis,
8 Lower arm flexor muscles

4) Dorsal Region – Transverse Sectional Plane

Transducer position: The transducer is positioned superior to the distal humerus in the transverse sectional plane. The transducer is moved from the upper distal third of the humerus toward the elbow and is positioned superior to the olecranon fossa.

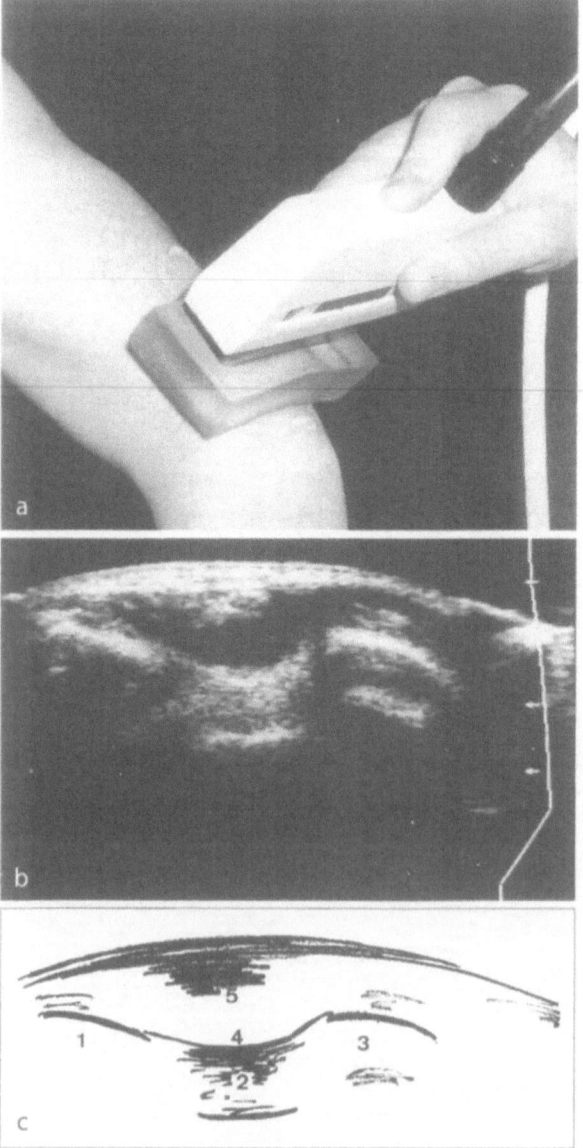

Fig. 9.4 a–c. Elbow joint, dorsal region – transverse sectional plane.
b, c.
1 Medial epicondyle,
2 Olecranon fossa,
3 Lateral epicondyle,
4 Joint capsule,
5 M. triceps brachii

5) Dorsal Region – Longitudinal Sectional Plane

Transducer position: The transducer position is superior to the distal humerus in the longitudinal sectional plane. The transducer is positioned along the long axis of the humeral shaft superior to the distal humerus and the olecranon. The osseous guiding structure of the distal humerus is positioned parallel to the upper monitor edge.

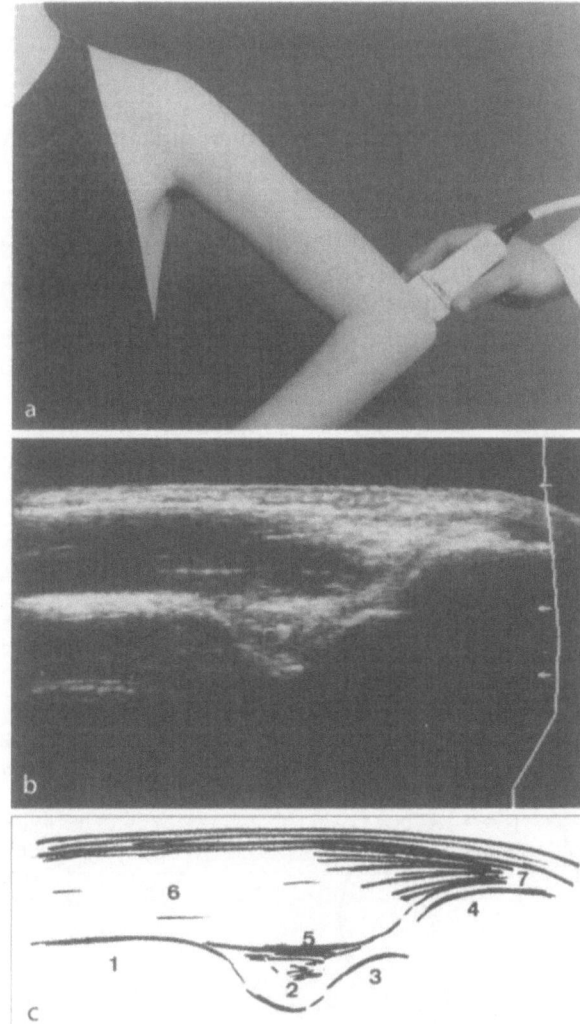

Fig. 9.5 a–c. Elbow joint, dorsal region – longitudinal sectional plane.
b, c.
1 Humeral shaft,
2 Olecranon fossa,
3 Humeral trochlea,
4 Olecranon,
5 Joint capsule,
6 M. triceps brachii,
7 Insertion zone of triceps tendon

9.2 Lateral Epicondylitis (Tennis elbow)

9.2.1 Definition

Tendon degeneration (endogenous) accompanied by simultaneous overexertion (exogenous) of the muscle insertion of the hand extensors at the lateral epicondyle.

9.2.2 Pathology

Tendon degeneration is characterized by loss of elasticity caused by fluid deficiency and sclerosis. Force transmission from the muscle to the bone is not absorbed optimally anymore by the insertion zone. This development is a normal aging process enforced by overexertion and microtrauma. Metabolic products are not transported away fast enough anymore; they are deposited in the insertion tissue causing necroses, calcium and salt deposits and ossification.

Histological studies show fatty degeneration and fibrous loosening at the tendon insertion area rather than inflammatory changes (contrary to the nomenclature: -itis).

Primary myogelosis of the extensor muscles is caused by overexertion. It is characterized by palpatory pain upon pressure of the epicondyle and functional flexion and extension pain. This includes hand and digit extension against a resistance.

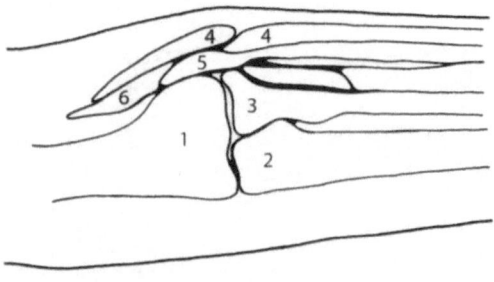

Fig. 9.6. Anatomy of the elbow – longitudinal sectional plane

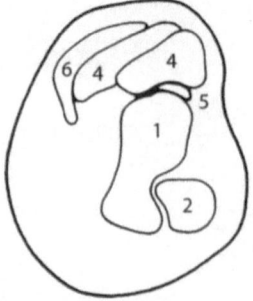

Fig. 9.6. and **9.7.**
1 Humerus,
2 Ulna,
3 Radius,
4 M. ext. carpi rad. longus/brevis,
5 Shared extensor tendon,
6 M. brachioradialis

Fig. 9.7. Anatomy of the elbow – transverse sectional plane

Secondary myogelosis is also characterized by a state of muscle tension. However, it develops as a protection measure against different provoking pathomechanisms, which include trauma of the joint cartilage (weight lifters, boxers, gymnasts) and overexertion of the annular ligament (lig. annulare radii). Secondary myogelosis is often diagnosed in fencers, tennis players and wrestlers, but it can also be found in other sports and in cases of physical stress (for instance in bricklayers).

9.2.3 Clinical Diagnosis

Lateral epicondylar pain upon palpation without visual signs of inflammation. Symptoms increase when the hand is extended against resistance. Pain inhibits the extension of the elbow joint.

X-ray: In most cases no pathological changes, such as insertion calcifications.

Ultrasound: Modified ventral humero radial longitudinal sectional plane: Possibly increased homogeneity (scarring) or lowered homogeneity (adipose) of the tendon tissue.

9.2.4 Conservative and Surgical Therapy

Conservative
Physical Therapy: Ultrasound treatment, cryotherapy
1. Physiotherapy
2. Transverse friction and stretching
3. Anti-inflammatory medication
4. Bandages
5. Injections, possibly including cortisone
6. Immobilization and avoidance of the cause of pain.

Surgical
The following are established surgical procedures
1. Hohmann's procedure: Dissection of the muscle at the point of origin
2. Wilhelm's procedure: Denervation.

9.2.5 ESWT

1) Prerequisites

1. Patient information given and written consent obtained from the patient.
2. Exclusion of contraindications.

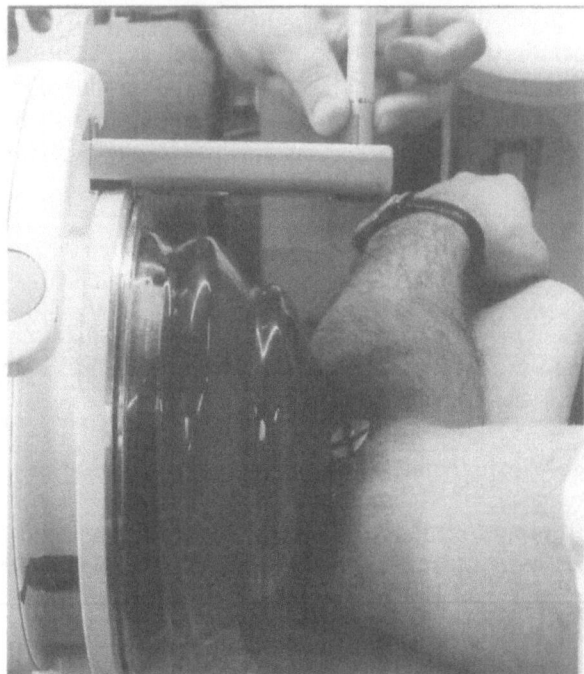

Fig. 9.8. ESWT treatment of lateral epicondylitis using patient feedback

2) Anesthesia

Low-energy therapy normally does not require anesthesia. If necessary, local anesthetic is recommended.

3) Positioning and Localization

Positioning depends on the region to be treated and on the type of localization system in use. The patient is sitting or in the supine position.

Patient Feedback (Laser Pointer)

Positioning: The patient is sitting or lying in the supine position on a treatment table with the arm pronated.

Localization: Palpation and marking of the ROI (X). Determination of the lateral coupling point of the therapy head (O). Matching of the laser pointer and the point (X) of maximum pain (=definition of penetration depth).

Ultrasound

Positioning: The patient is sitting or lying in the supine position on a treatment table with the arm pronated.

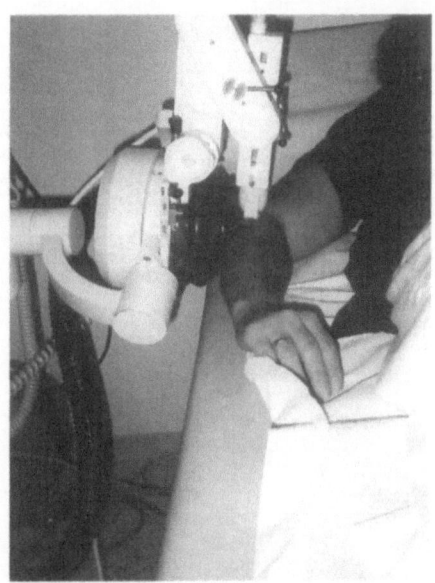

Fig. 9.9. ESWT treatment of lateral epicondylitis using ultrasound localization

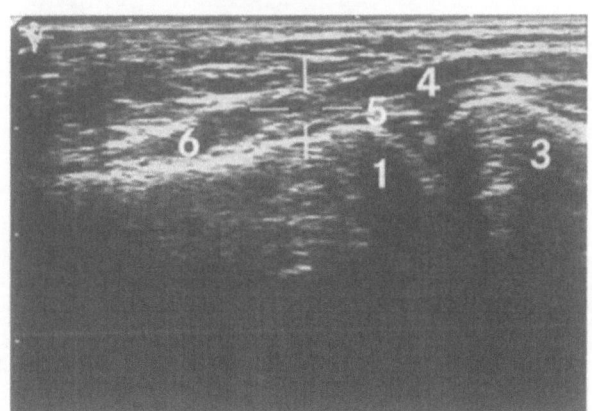

Fig. 9.10. Lateral epicondylitis, modified humero-radial longitudinal sectional plane

Localization: Modified humero radial longitudinal sectional plane (c.f. Fig. 9.10).

View of the humeral shaft, the lateral epicondyle (1), the radius head (3) and the extensor muscles.

The ultrasound transducer allows one to target the insertion point of the tendon accurately.

X-ray

Positioning: The patient is required to be on a treatment table, when using X-rays.

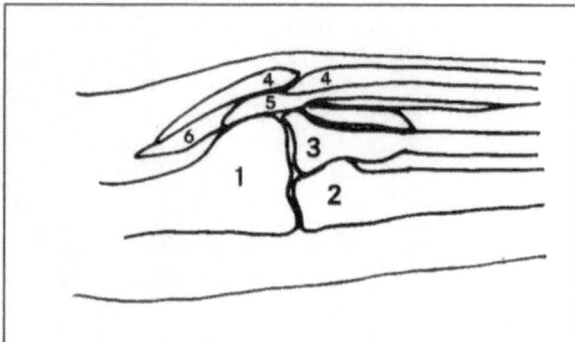

Fig. 9.11. Anatomy of the elbow – longitudinal sectional plane
1 Humerus,
2 Ulna,
3 Radius,
4 M. ext. carpi rad. longus/brevis,
5 Shared extensor tendon,
6 M. brachioradialis

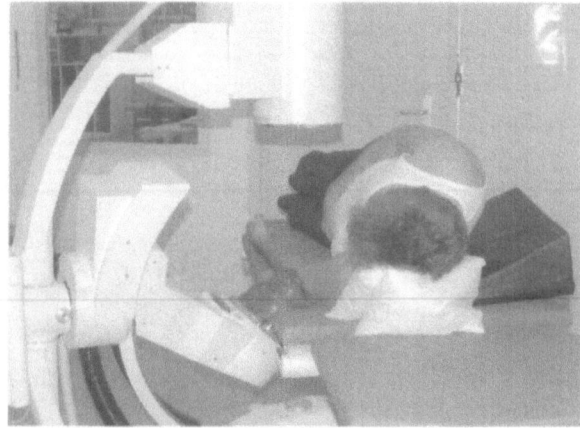

Fig. 9.12. ESWT treatment of lateral epicondylitis using X-rays

- Supine position.
- Treatment side is facing the therapy head.
- Arm is slightly flexed and suspended over the table cutout. It is fixed by the arm support.
- Therapy head is moved from lateral to the epicondyle.
- Marking the point of maximum pain and adjusting the localization system or the laser pointer.

When using X-rays, the patient should be protected against unnecessary exposure to radiation!

Localization: Imaging of lateral epicondyle, first in anterior-posterior sectional plane. After aligning the crosshairs with the ROI, tilting of the C-arm and adjusting in the second sectional plane.

4) Coupling

Tangential Coupling Using the Two Point Method (TPM)
1. Elbow slightly flexed.
2. Identify and mark painful pressure point on the lateral epicondyle. This point marks the location for shock wave application.
3. The depth of the tender point, either close to the surface or deeper in the tissue is determined by palpation. The depth for the shock wave focus is determined accordingly.
4. Coupling of the therapy head to the anterior mark.
   Remember the coupling gel!
5. Take precise aim at the tendon insertion with the ultrasound transducer (ventral coupling with longitudinal and horizontal imaging).

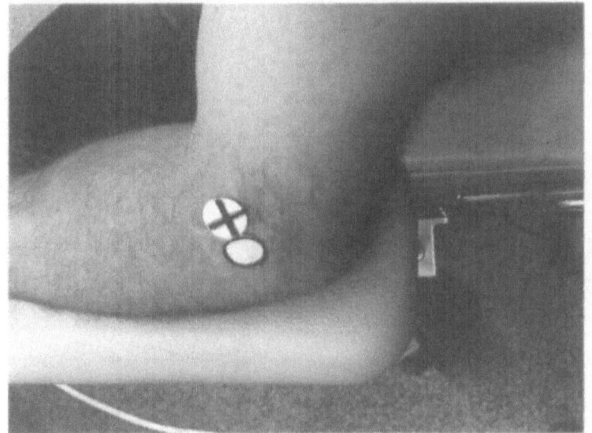 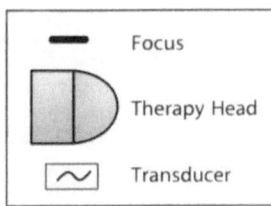

Fig. 9.13. Lateral epicondylitis: Tangential coupling (TPM)

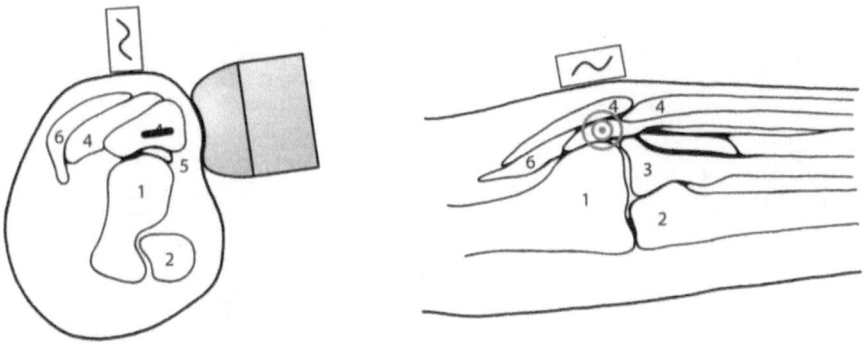

a   b

Fig. 9.14 a, b. Tangential coupling (TPM). view from ventral (**a**) and from lateral (**b**)

Direct Coupling Using the One Point Method (OPM):
1. Position patient as above.
2. Identify and mark tender point = entrance point of the shock wave (X).

---

5) Treatment Regimen

The therapy head is positioned by precise X-ray, ultrasound or laser pointer localization. During treatment correct positioning must be controlled.

The procedure should be verified at least twice when using X-rays for localization. Ultrasound allows continuous, real-time localization control.

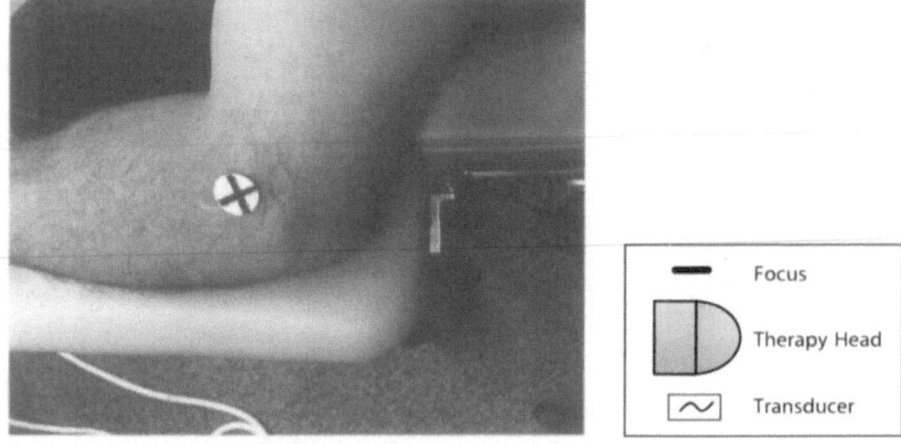

Focus

Therapy Head

Transducer

Fig. 9.15. Lateral epicondylitis, direct coupling (OPM)

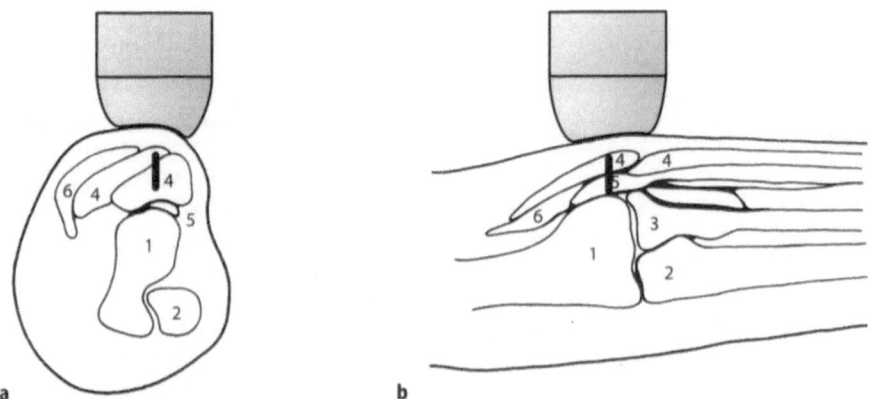

Fig. 9.16. Direct coupling (OPM), view from ventral (**a**) and lateral (**b**)

The number of pulses and number of therapy sessions applied are determined by the therapy concept of the treating physician. The required energy level can be reached through gradually increasing the energy level depending on the patient's sensitivity to pain and the reaction to analgesic effects caused by shock wave application.

The recommended treatment regimen is 3 treatment sessions of 1500–2000 shock waves per session. The cumulated positive energy flux density summed up over all sessions should be approximately 1300 mJ/mm^2.

9.3 Medial Epicondylitis (Golfer's Elbow)

9.3.1 Definition

Tendon degeneration (endogenous) accompanied by simultaneous overexertion (exogenous) of the muscle insertion of the hand flexors at the medial epicondyle.

9.3.2 Pathology

Tendon degeneration is characterized by loss of elasticity caused by fluid deficiency and sclerosis. Force transmission from the muscle to the bone is not absorbed optimally anymore by the insertion zone. This development is a normal aging process enforced by overexertion and microtrauma. Metabolic products are not transported away fast enough anymore; they are deposited in the insertion tissue causing necroses, calcium and salt deposits and ossification.

Histological studies show fatty degeneration and fibrous loosening at the tendon insertion area rather than inflammatory changes (contrary to the nomenclature: -itis).

Primary myogelosis of the extensor muscles is caused by overexertion. It is characterized by palpatory pain upon pressure of the epicondyle and functional flexion and extension pain. This includes hand and digit extension against a resistance.

Secondary myogelosis is also characterized by a state of muscle tension which develops as a protection measure against other provoking pathomechanisms, which include trauma of the joint cartilage (weight lifters, boxers, gymnasts) and overexertion of the annular ligament (lig. annulare radii). Golfers are often diagnosed with this condition, but it can also occur performing other sports or physical activities.

9.3.3 Clinical Diagnosis

Medial epicondylar pain upon palpation with lack of visual signs of inflammation. Symptoms increase when the hand is flexed against resistance. Pain impedes the bending of the elbow joint.

X-ray: Mostly no pathological changes, such as insertion calcifications.

Ultrasound: Modified ventral humero ulnar longitudinal sectional plane: Possibly increased homogeneity (scarring) or lowered homogeneity (adipose) of the tendon tissue.

9.3.4 Conservative and Surgical Therapy

Conservative
1. Physical therapy: Ultrasound treatment, cryotherapy
2. Active physiotherapy
3. Transverse resistance and stretching
4. Anti-inflammatory medication
5. Bandages
6. Cortisone injections
7. Immobilization and avoidance of the cause for pain.

Surgical
The following are established procedures:
1. Hohmann's procedure: Dissection of the muscle at the point of origin
2. Wilhelm's procedure: Denervation.

9.3.5 ESWT

1) Prerequisites

1. Patient information given and written consent form obtained from the patient.
2. Exclusion of contraindications.

2) Anesthesia

Caution: Anesthesia should not be administered when treating golfer's elbow. This will insure that the N. ulnaris (ulnar nerve), which is in close anatomical proximity is not accidentally in the focal range (patient feedback).

3) Positioning and Localization

Positioning depends on the region to be treated and the locating system being utilized. The treatment can be performed with the patient sitting or in the supine position.

Patient Feedback (Laser Pointer)

Positioning: The patient is sitting or lying in the supine position on a treatment table with the arm supinated.

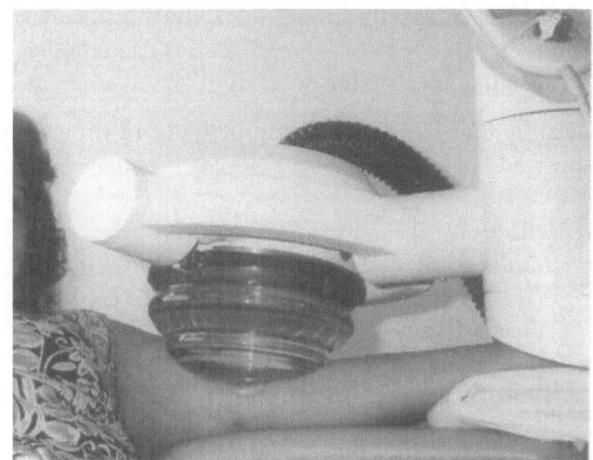

Fig. 9.17. ESWT treatment of medial epicondylitis using patient feedback

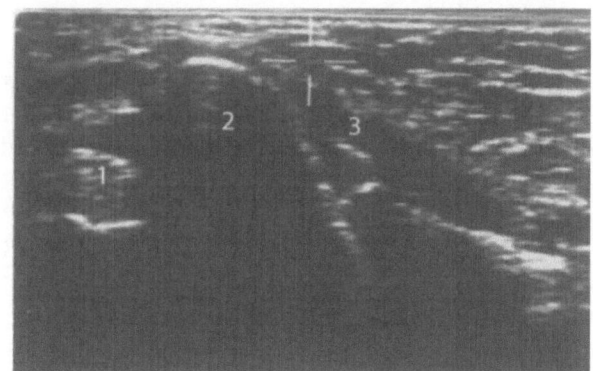

Fig. 9.18. Modified medial longitudinal sectional plane for medial epicondylitis.
1 N. ulnaris,
2 Medial epicondyle,
3 M. flexor carpi radialis longus and M. digitorum

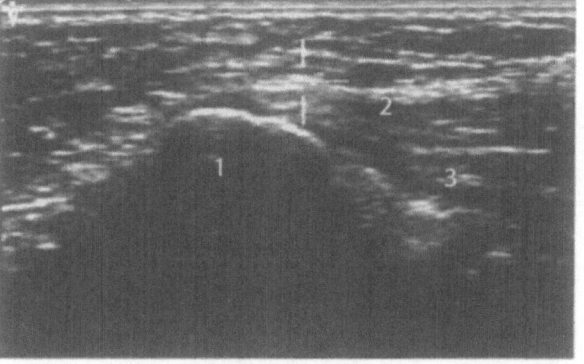

Fig. 9.19. Modified medial transverse sectional plane for medial epicondylitis
1 Humerus,
2 M. flexor carpi radialis longus,
3 M. flexor digitorum

Localization: Palpation and marking of the ROI (X). Determination of the lateral coupling point for the therapy head (O). Matching of the laser pointer and the point (X) of maximum pain (=definition of maximum penetration depth).

Ultrasound Localization

Positioning: Patient is sitting or lying in the supine position with the arm supinated.

Localization:
■ Modified medial longitudinal sectional plane
View of the humeral shaft, medial epicondyle, flexor muscles and the N. ulnaris (transverse sectional plane).
■ Modified medial transverse sectional plane:
View of the medial epicondyle, flexor muscles and the N. ulnaris.

X-ray Localization

Positioning:
■ Patient is supine
■ Treatment side is facing the therapy head
■ Elbow is extended on the arm rest and in the supine position
■ Therapy head is moved from medial to the medial epicondyle
■ Marking of the point of maximum pain and adjustment of the localization systems and laser pointer, respectively.

Do not forget to protect patient from radiation when using X-rays!

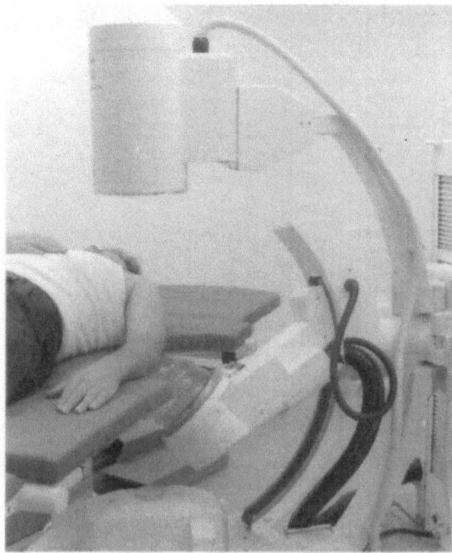

Fig. 9.20. ESWT treatment of medial epicondylitis using X-rays

Localization: Imaging of the medial epicondyle, first in antero-posterior projection. After aligning the crosshairs with the ROI, tilting of C-arm and adjusting in the second sectional plane.

4) Coupling

Tangential Coupling using the Two Point Method (TPM):
1. Elbow is extended and supinated.
2. Identify and mark painful pressure point at the medial epicondyle. This point marks the location of shock wave application.
3. The depth of the tender point is determined by palpation, either on the surface or deeper in the tissue. The depth for the shock wave focus is determined accordingly.
4. Coupling of the therapy head to the anterior mark.
   Remember the coupling gel!
5. Take precise aim at the tendon insertion with the ultrasound transducer (ventral coupling with longitudinal and horizontal imaging).

Direct Coupling using the One Point Method (OPM)
1. Position patient as above.
2. Identify and mark the point of maximum pain=entrance point of the shock wave (X).

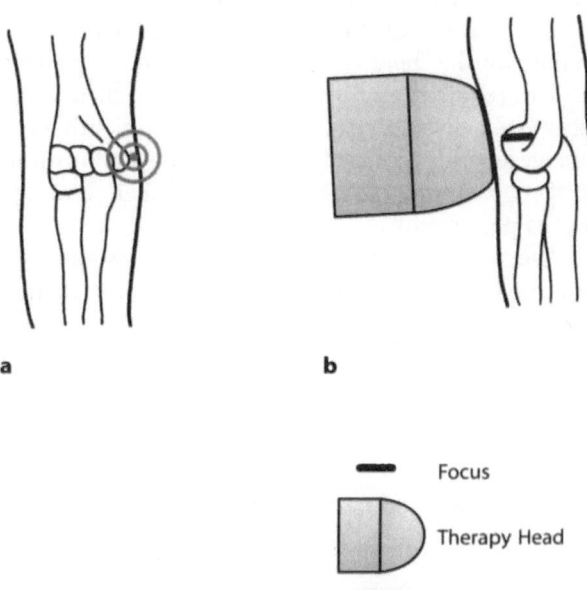

▬	Focus
◖⊃	Therapy Head
⊡	Transducer

Fig. 9.21 a, b. Tangential coupling (TPM), view from lateral (**a**), and ventral (**b**)

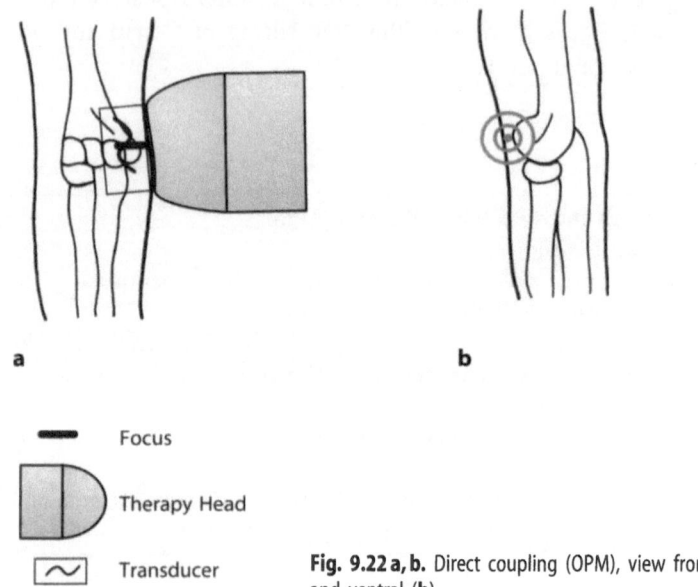

a   b

■—   Focus

        Therapy Head

[~]   Transducer   **Fig. 9.22 a, b.** Direct coupling (OPM), view from medial (**a**), and ventral (**b**)

5) Treatment Regimen

The therapy head is positioned by precise X-ray, ultrasound or laser pointer localization. During treatment correct positioning must be controlled.

The procedure should be verified at least twice, when using X-rays for localization. Ultrasound provides continuous, real-time localization control.

The number of pulses and number of therapy sessions applied are determined by the therapy concept of the treating physician.

The required energy level can be reached by gradually increasing the energy level depending on the patient's sensitivity to pain and the reaction to analgesic effects caused by shock wave application.

The recommended treatment regimen is 3 sessions of 1500–2000 shock waves per session.

The recommended cumulated positive energy flux density per treatment summed up over all sessions should be approximately 1300 mJ/mm^2.

9.4 Results

Table 9.2. Recently published results of shock wave therapy of lateral epicondylitis

Author	Patients	Follow-up (months)	Success rate %	Energy
Hammer 2000	19	6 months	63	3×3000, 0.12 mJ/mm^2
Krischek 1998	30	24 weeks	60	2×500, 0.08 mJ/mm^2
Haist 2000	1098	12 months	72	2.9×1500, 0.08-0.23 mJ/mm^2
Rompe 1997	50	52 weeks	52	3×1000, 0.08 mJ/mm^2
Rompe 1996	50	24 weeks	48	3×1000, 0.08 mJ/mm^2
Dahmen 1995	46	12.1 months	69.6	1-10×0.08 mJ/mm^2
Levitt 2000	20	12 months	85	2×1000, 16 KV

Table 9.3. Recently published results of shock wave therapy of medial epicondylitis

Author	Patients	Follow-up	Success rate %	Energy
Krischek 1998	30	24 weeks	27	2×500, 0.08 mJ/mm^2
Haist 2000	308	12 months	66	2.9×1500, 0.08-0.23 mJ/mm^2

Various study groups reported that the application of ESWT for the treatment of epicondylitis can yield good results. First, all treatments were done at low energy. With numerous repeated treatments, as performed by In Dahmen et al., a significant improvement was achieved in 70% of patients with lateral epicondylitis. These patients were free from complaints or with clearly reduced pain after ESWT. Although the study by Dahmen et al. does not comply with the necessary biometric standards, its clinical evidence proves that ESWT has the potential to become a true alternative to surgery.

In improved studies of better quality, as performed by Rompe and by Krischek et al., these first results where confirmed. The interval between sessions does not seem to have a large influence on the results.

In a feasibility study Levitt et al. achieved complete or significant freedom from complaints in up to 85% of patients after one to two treatment sessions. This study has been expanded to a multicenter, randomized, placebo-controlled study with 300 patients to be enrolled.

Positive effects shown in numerous studies after 3 months or after a shorter interval also remain stable over a longer period. Thus, Rompe et al. observed very good and good results in 52% of the cases after 52 weeks in a population of 50 patients. In these cases 3 times 1000 shock waves were applied with an average energy flux density of 0.08 mJ/mm^2.

None of the authors reported serious complications as long as general recommendations for shock wave therapy were observed. Occasionally slight skin reddening was observed.

Contrary to results for lateral epicondylitis, the results of shock wave therapy for medial epicondylitis (Golfer's elbow) yield a less uniform picture with very different results.

The group of Haist et al. reports good results in over 300 patients for instance. These were treated, sometimes in several sessions, with different energy flux densities leading to good to very good treatment results in 66% of the patients after one year.

Contrary to these results is the work by Krischek et al. Using a standardized protocol, they observed good and very good treatment results in only 27% of the patients. They treated 30 patients at low energy in two sessions with 500 shock waves using 0.08 mJ/mm^2.

Again these authors observed no serious complications when treating medial epicondylitis.

10 Foot and Ankle Joint

10.1 Standard Ultrasound Sectional Planes of the Ankle Joint and Foot

10.1.1 Standard Indications and Findings

Ultrasound of the ankle joint can detect intraarticular volume increases in the presence of effusion or synovialitis, as well as periarticular soft tissue changes. The most frequent indications for ultrasound examination of this joint are injuries and disorders of the Achilles tendon. Ultrasound can offer decisive additional information for the conservative treatment of Achilles tendon rupture. In addition to ruptures, inflammatory-degenerative changes of the Achilles tendon, such as Achillodynia and paracalcaneal inflammatory changes, can be sonographically detected. Changes of the plantar fasciitis following treatment with ESWT can be sonographically detected. A rupture of the anterior syndesmosis reveals a typical sonomorphological change. Table 10.1 lists general and rare indications for ultrasound examination of the ankle.

Table 10.1. Typical indications and findings

- Osseous destruction/lesion
- Osteophyte
- Plantar fasciitis – heel spur
- Free arthrophytes
- Chondromatosis
- Intraarticular volume increase
- Bursitis
- Peritendineum changes
- Achillodynia
- Achilles tendon rupture
- Anterior syndesmosis changes
- Fibular band apparatus changes
- Inflammatory-rheumatoid diseases
- Fracture
- Tumor

10.1.2 Examination

For ultrasound examination of the ventral standard sectional planes, the patient is in the supine position. For ultrasound examination of the dorsal sectional planes, the patient is in the prone position. The patient's feet, when in prone position, should extend beyond the examination table with the knee and hip joints in a neutral-zero position. The physician has a very good view of the gliding motion of the Achilles tendon using dynamic examination in the dorsal longitudinal sectional plane.

10.1.3 Standard Sectional Planes

For ultrasound images, the ankle joint is examined in three regions using five standard sectional planes. In the ventral and dorsal longitudinal region, the joint is imaged in two sectional planes that are almost vertical to each other; a diagonal sectional plane is used for the lateral region.

- Ventral region
  - Transverse sectional plane
  - Longitudinal sectional plane
- Lateral region
  - Diagonal sectional plane
- Dorsal region
  - Transverse sectional plane
  - Longitudinal sectional plane.
- Plantar region
  - Transverse sectional plane
  - Longitudinal sectional plane

If the finding is non-pathological, verification of two standard sectional planes is recommended:

- Ventral region
  - Longitudinal sectional plane
- Dorsal Region
  - Longitudinal sectional plane.

In the following, for each standard sectional plane, the transducer position, a standard ultrasound image and an explanatory diagram are presented.

1) Ventral Region – Transverse Sectional Plane

Transducer position: The transducer is positioned near the talus in the transverse sectional plane. The transducer is guided from the distal third of
the tibia towards the ankle and positioned superior to the trochlea of the
talus.

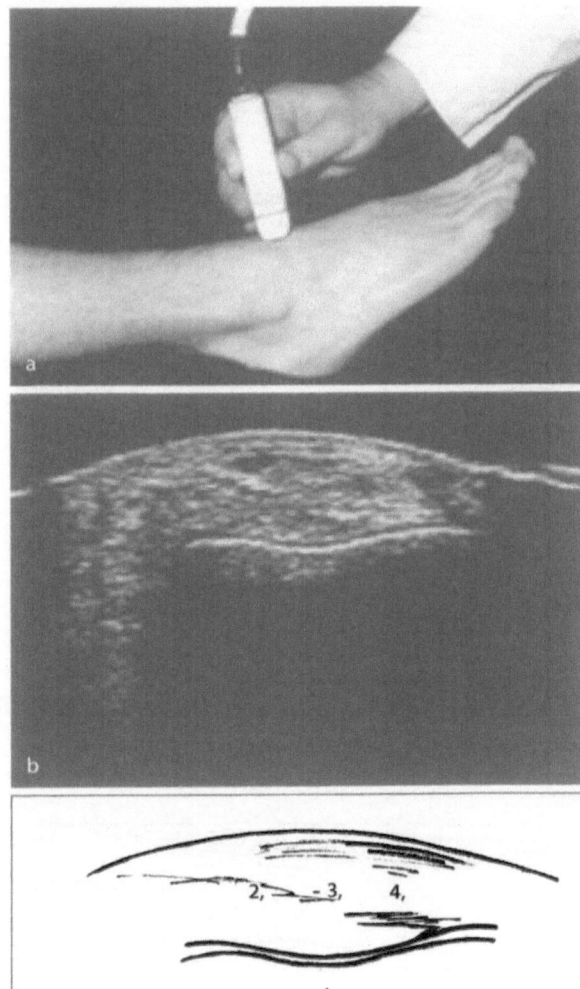

Fig. 10.1 a–c. Foot-ankle joint,
ventral region – transverse sectional plane.
b, c.
1 Talus
2 M. tibialis anterior
3 M. extensor hallucis longus
4 M. extensor digitorum longus

2) Ventral Region – Longitudinal Sectional Plane

Transducer position: The transducer is positioned in the longitudinal sectional plane superior to the distal tibia and the talus. The transducer position corresponds to the long axis of the distal tibia. The ventral cortex of the tibia should be parallel to the upper edge of the monitor.

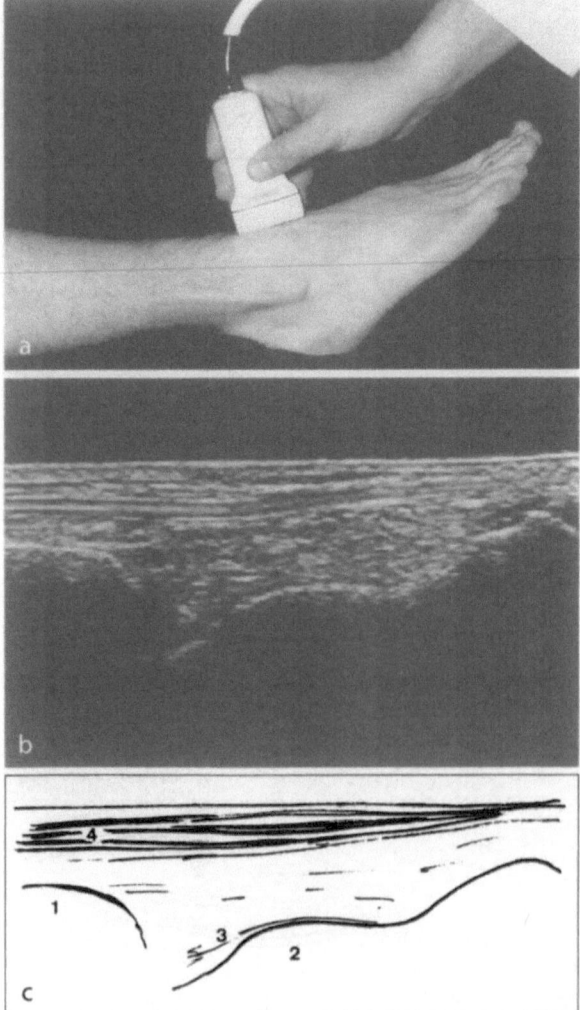

Fig. 10.2 a–c. Foot-ankle joint, ventral region – longitudinal sectional plane.
b, c.
1 Tibia
2 Talus
3 Joint capsule
4 M. extensor hallucis longus

3) Lateral Region – Diagonal Sectional Plane

Transducer position: The transducer is positioned longitudinally superior to the anterior syndesmosis.

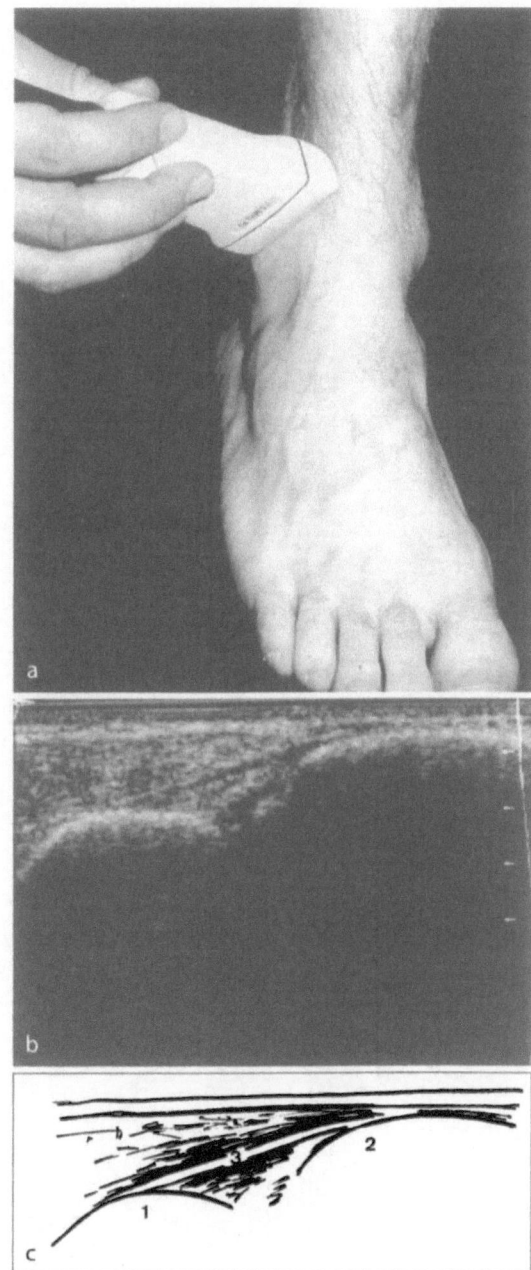

Fig. 10.3 a–c. Foot-ankle joint, lateral region – diagonal sectional plane.
b, c.
1 Tibia
2 Fibula
3 Anterior syndesmosis

4) Dorsal Region – Transverse Sectional Plane

Transducer position: The transducer is positioned superior to the Achilles tendon in the transverse plane. The transducer is moved from the distal third of the tibia toward the ankle and is positioned superior to the Achilles tendon.

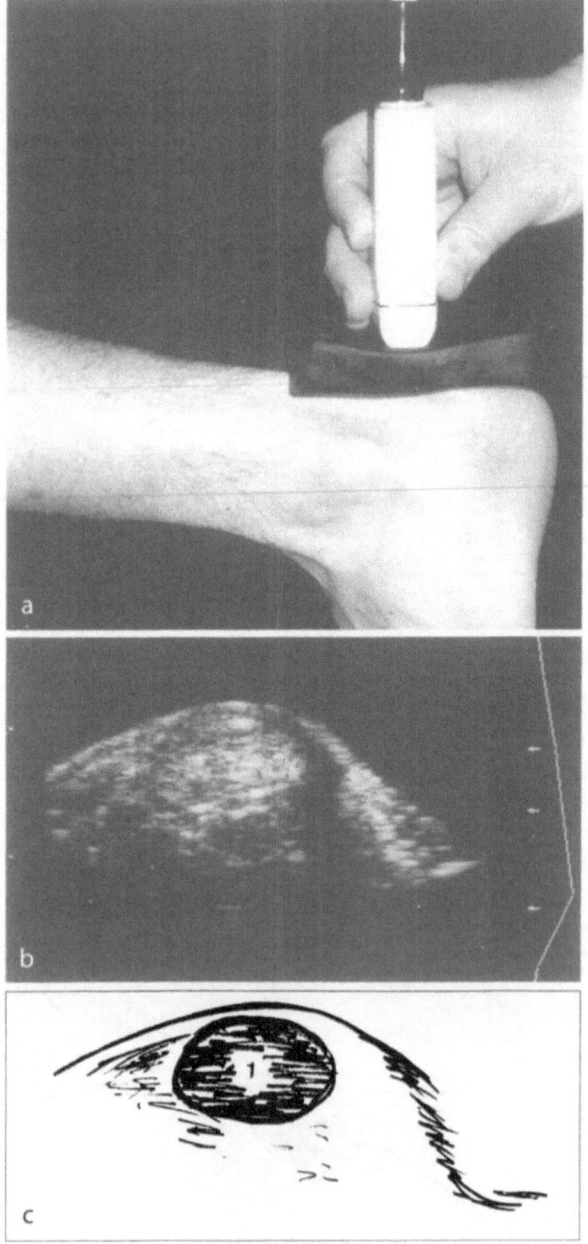

Fig. 10.4 a–c. Foot-ankle joint, dorsal region–transverse plane.
b, c.
1 Achilles tendon

5) Dorsal Region – Longitudinal Sectional Plane

Transducer position: The transducer is positioned longitudinally, superior to the Achilles tendon, the dorsal region of the distal tibia, the talus and the calcaneus.

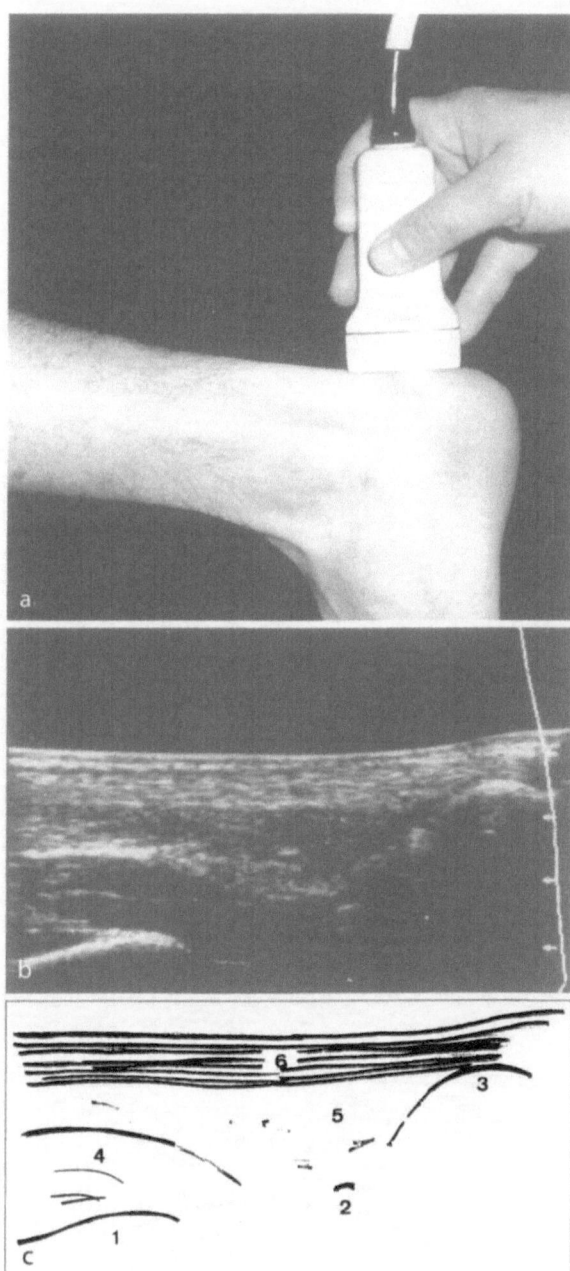

Fig. 10.5 a–c. Foot-ankle joint, dorsal region – longitudinal plane.
b, c.
1 Tibia
2 Talus
3 Calcaneus
4 M. flexor hallucis longus
5 Kager's triangle
6 Achilles tendon

6) Plantar Region – Transverse Sectional Plane

Transducer position: The transducer is positioned transverse to the sole, and is moved from the heel to the tarsus. The echogenic cortical substance of the calcaneus appears osseous. The plantar fascia appears echogenic when viewed from an orthograde angle versus echo poor when viewed from tangential angle.

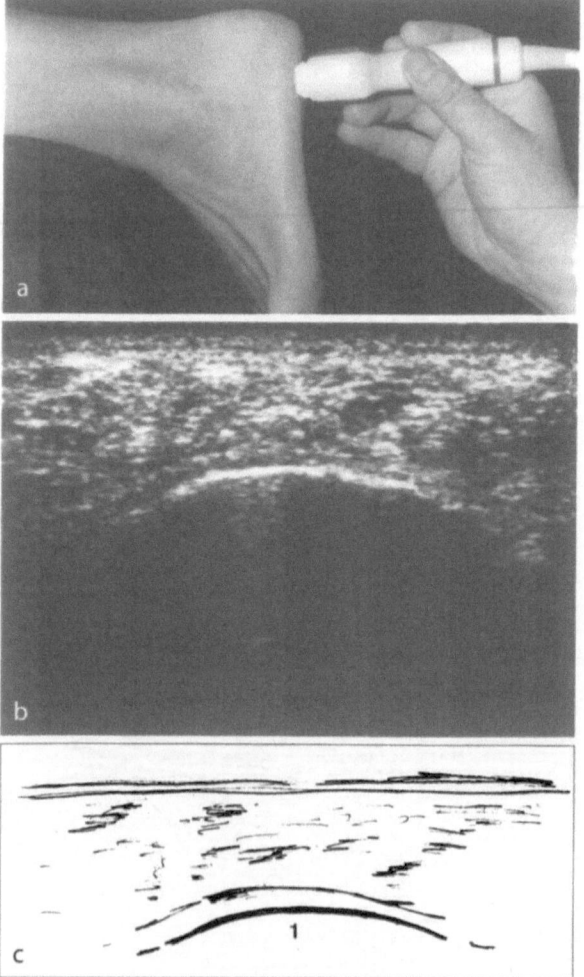

Fig. 10.6 a–c. Foot-ankle joint, plantar region – transverse plane
b, c.
1 Calcaneus

7) Plantar Region – Longitudinal Sectional Plane

Transducer position: The transducer is positioned longitudinally to the sole, and is moved from the heel to the tarsus. The echogenic cortical substance of the calcaneus and possibly the tarsus bone appear osseous. The plantar fascia appears echogenic with an echo poor triangle shaped insertion area at an orthograde sound angle.

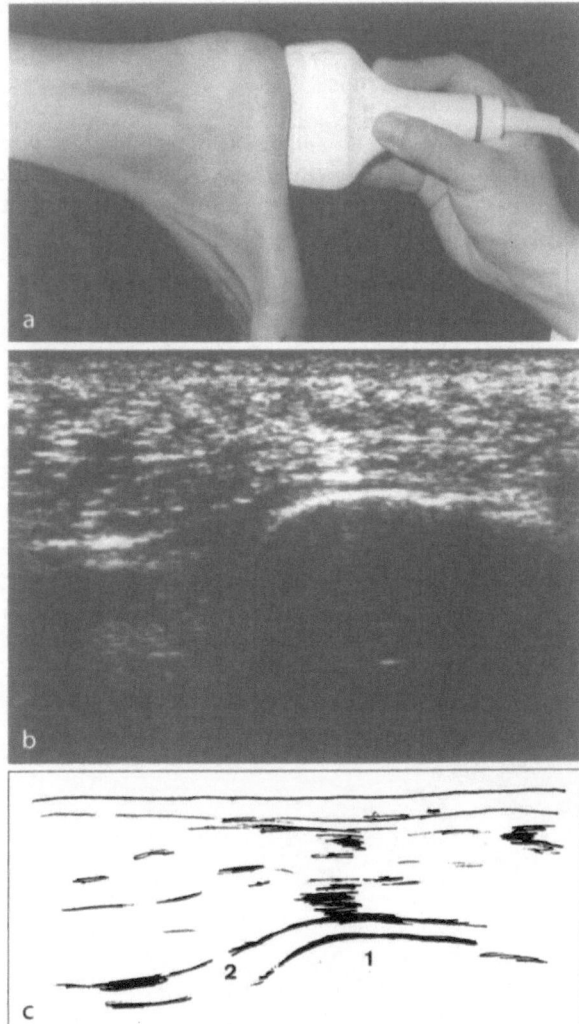

Fig. 10.7 a–c. Foot-ankle joint, plantar region – longitudinal plane
b, c.
1 Calcaneus,
2 Plantar fascia

10.2 Plantar Fasciitis (Plantar Heel Spur)

10.2.1 Definition

Spur-like, reactive bone formation at the calcaneal medial tubercle.

10.2.2 Pathology

This disease is caused by chronic overloading of the small muscles of the foot (M. abductor hallucis, M. flexor digitorum brevis) and by plantar aponeurosis (plantar fasciitis) in the case of malformations of the foot (pathological change of the talo-calcaneal angle) such as pes valgus/flat foot. This leads to an increased direct load on the medial surface of the calcaneus. Indirectly the pathologically changed talo-calcaneal angle of the longitudinal vault of the foot causes increased tension in the insertion area.

10.2.3 Clinical Diagnosis

Women over 40 years of age, overweight patients, and those patients, whose work requires long periods of standing, most frequently suffer from plantar fasciitis, often accompanied by the development of a heel spur, resulting from pes valgus/flat foot in the majority of cases.

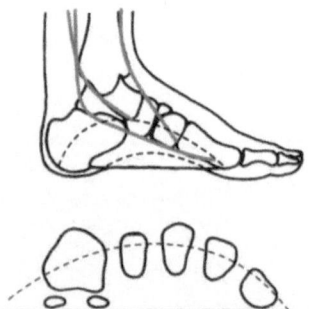

Fig. 10.8. Anatomy, normal foot

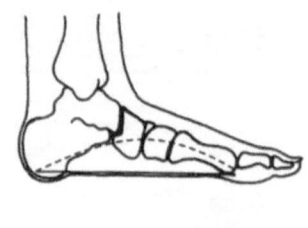

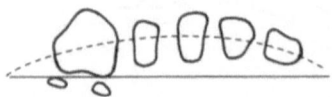

Fig. 10.9. Anatomy, pes valgus/flat foot

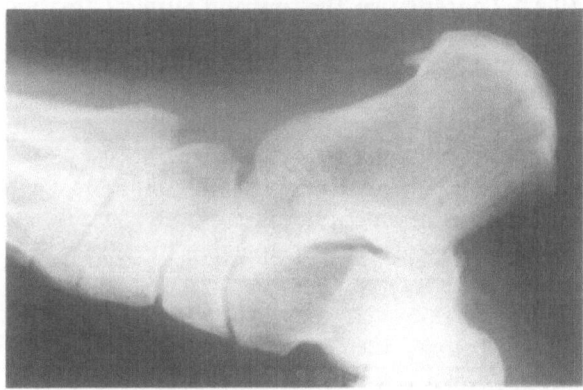

Fig. 10.10. Lateral X-ray image of a heel. A typical plantar heel spur is shown. The X-ray image corresponds to the ultrasound image (Fig. 10.11)

Pain may be localized (medial heel) or distributed over the entire heel and radiate to the metatarsus and forefoot.

Gait is significantly influenced, the heel is often non weight bearing, causing patients to walk on their forefoot.

X-ray: Exostosis at the calcaneal tuberosity.

Calcification may develop later than clinical symptoms since the principal disease is insertional tendinitis.

Ultrasound: Plantar fasciitis affects the insertion region at the calcaneus. A thickening of the tendon with a corresponding change in homogeneity can be observed in most cases. The heel spur is imaged as an osseous prominence above level with sound shadow.

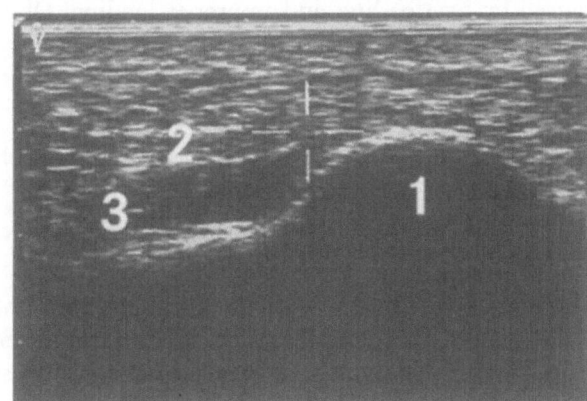

Fig. 10.11. Plantar region, longitudinal sectional plane of the medial calcaneus section.
1 Calcaneus
2 Plantar fascia
3 Short foot muscles

10.2.4 Conservative Therapy and Surgical Treatment

Conservative
1. Use of orthotics for plantar arch support and of orthopedic cushions to reduce load on the heel spur area.
2. Support of plantar arch and supination of the heel. Excessive supination must be avoided, because of the risk of supination trauma with possible damage to fibular ligaments.
3. Antiphlogistic medication.
4. Stimulating X-ray irradiation.
5. Local injections, possibly with corticosteroid injections.

Surgical
1. Release of the plantar aponeurosis and resection of short muscles of the foot.
2. Excision of the heel spur.

Because of its generally poor outcome, surgery is considered to be the last method.

10.2.5 ESWT

Basic Observation when Treating Plantar Fasciitis (Heel Spur)

The goal of ESWT is the treatment of chronic inflammation of the tendon/fascia and not the fragmentation of the exostosis.

The exostosis is a secondary consequence of chronic fasciitis, which is caused by overloading and improper loading. Of all exostoses 90% are clinically asymptomatic and require no treatment.

1) Prerequisites

1. Patient information given and written consent obtained
2. Exclusion of contraindications.

2) Anesthesia

Therapy at the low energy level normally does not require anesthesia. If required, conduction anesthesia should be administered to the medial R. calcaneus (N. tibialis) region. A local anesthetic deposit should be administered under bony contact in the middle between the medial malleolus and the lower border of the calcaneus.

N. tibialis   R. calcaneus medialis

Fig. 10.12. N. tibialis, anatomical course in   R. calcaneus medialis
the foot

3) Positioning and Localization

Positioning depends on the region to be treated and the locating system used. The patient can be treated either sitting or in the prone or supine position.

Patient Feedback (Laser Pointer)

Positioning: The patient is either sitting or either in the prone or supine position on the treatment table. The foot and the distal third of the lower leg are freely movable.

Localization: Palpation and marking of the ROI (X). Determination of the medial coupling point for the therapy head (O). The laser pointer and the maximum pain pressure point (X) are aligned (= definition of penetration depth).

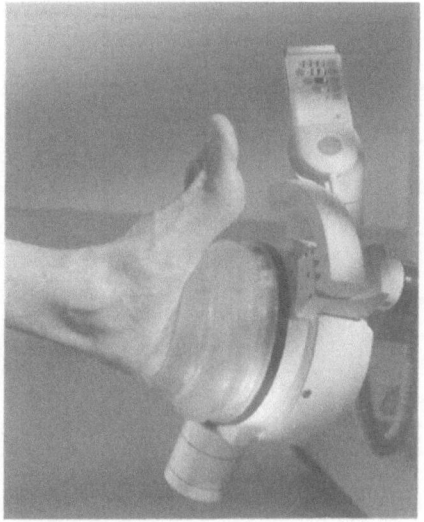

Fig. 10.13. ESWT treatment of the plantar fascia
with patient feedback localization

Ultrasound

Positioning: The patient is either sitting or either in the prone or supine position on the treatment table. The foot and the distal third of the lower leg are freely movable.

Localization: Plantar longitudinal plane. Image of the calcaneus and the plantar fascia. Focusing at the transition area between the calcaneus and plantar fascia.

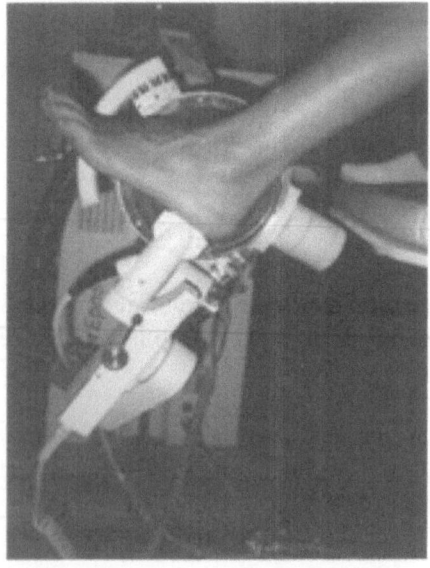

Fig. 10.14. ESWT treatment of plantar fasciitis with ultrasound localization (sitting/supine position)

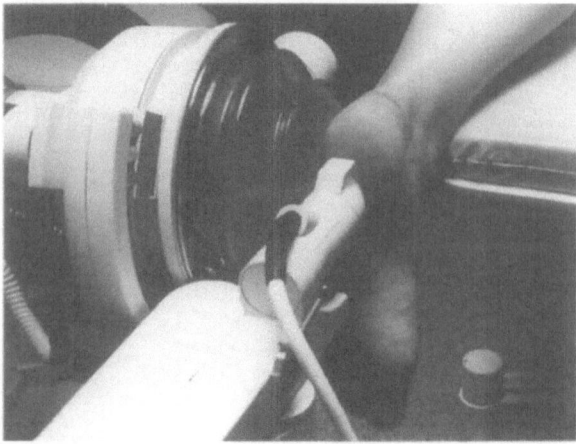

Fig. 10.15. ESWT treatment of plantar fasciitis with ultrasound (prone position)

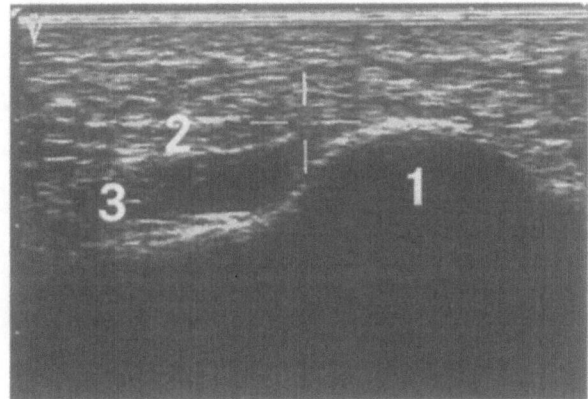

Fig. 10.16. Ultrasound focusing for the plantar fasciitis at the transition to the calcaneus (plantar region, longitudinal sectional plane).
1. Calcaneus
2. Plantar fascia
3. Short foot muscles

X-ray Localization

Positioning:
- The patient is either in the prone or supine position
- The therapy side is facing the therapy head (medial coupling preferred)
- Marking of the plantar point of maximum pain and adjustment of the C-arm.

Localization: In case of X-ray localization, protect patient from radiation!

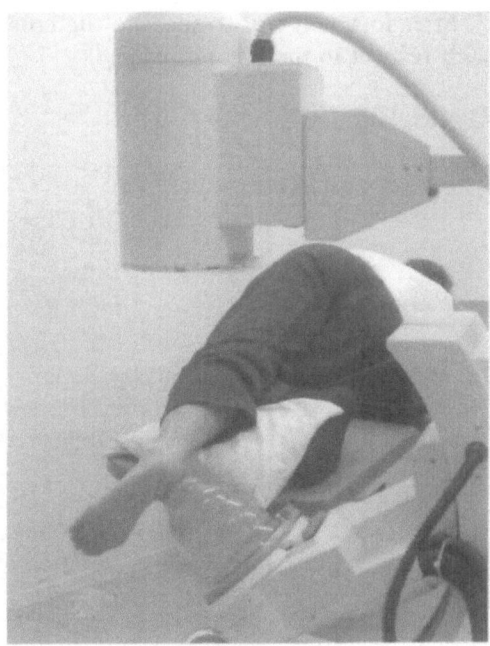

Fig. 10.17. ESWT treatment of plantar fasciitis with X-rays, tangential coupling

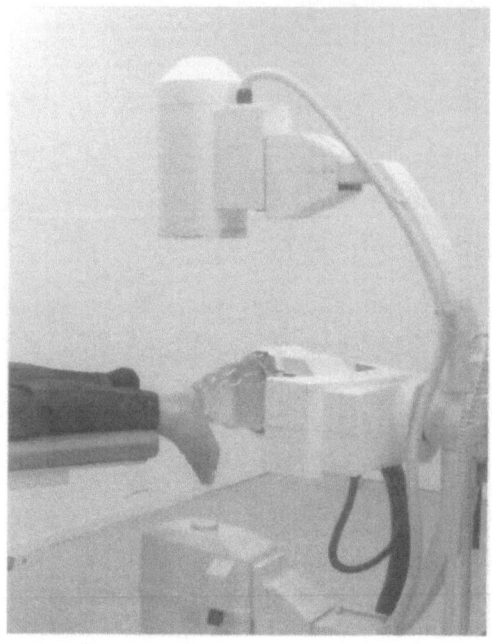

Fig. 10.18. ESWT treatment of plantar fasciitis with X-rays, direct coupling

Imaging of the calcaneus, with possible osseous heel spur, in the antero-posterior projection. Focusing at the heel spur or at the transition area of the plantar fascia.

After focusing with respect to the cross hairs, tilt of the C-arm and adjust with respect to the second plane.

4) Coupling

Tangential Coupling:
1. Internal or external rotation of the leg, the foot is parallel to therapy head.
2. Palpate, identify and mark the tender point on the plantar side of the calcaneus = *penetration depth of the shock wave.*
3. Identify the tender point on the medial/lateral side of the foot and mark intersection with calcaneus = *entrance point of the shock wave.*
4. To verify position, palpate the border of the calcaneus and the spur.
5. Couple the therapy head to the medial/lateral marking.
   Do not forget coupling gel!
6. Use the ultrasound transducer (coupling plantar with longitudinal and transverse imaging) for precise shock wave positioning.
7. Using the C-arm, align the crosshairs with the insertion calcification in 2 planes

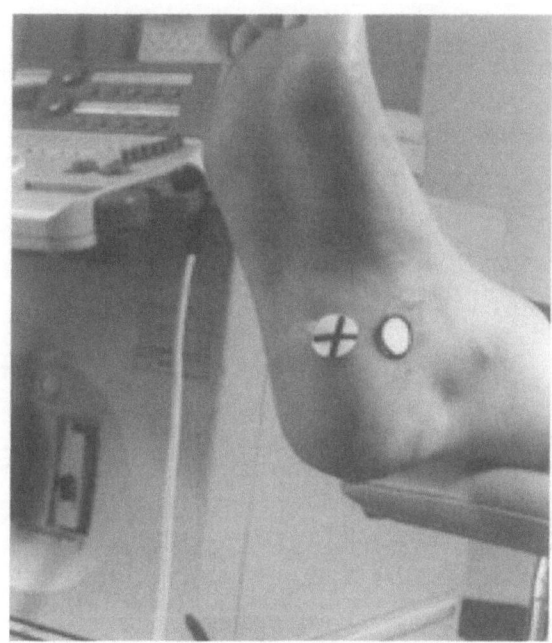

Fig. 10.19. Marking of the entrance point (O) of the shock wave for plantar fasciitis, tangential coupling

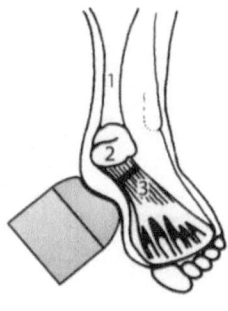

a

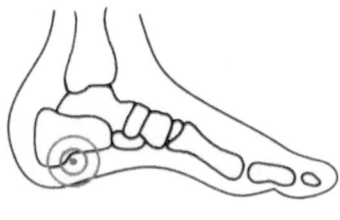

b

Fig. 10.20 a, b. View from different directions, tangential coupling for plantar fasciitis. **a** from medial, **b** from plantar

---

Direct Coupling (OPM)
1. Position patient as described above.
2. Identify and mark the tender point = entrance point of the shock wave (O).

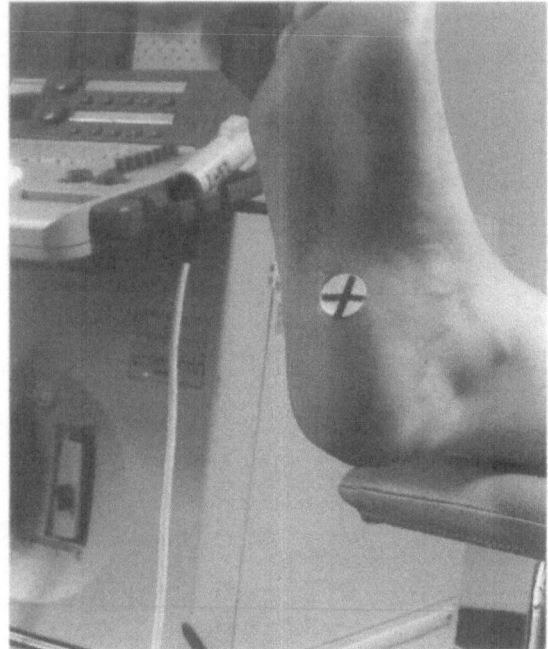

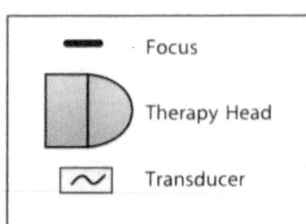

Focus

Therapy Head

Transducer

Fig. 10.21. Marking of the entrance point of the shock wave, direct coupling for plantar fasciitis

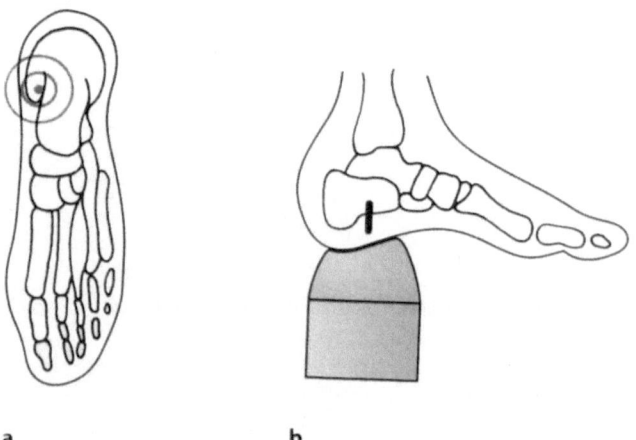

a   b

Fig. 10.22. View from different directions, direct coupling for plantar fasciitis

5) Treatment Regimen

Shock wave treatment should begin following precise X-ray or ultrasound localization. A low to mid-range energy setting is recommended for treatment of plantar fasciitis.

The localization should be verified at least twice, when using X-rays for localization. Ultrasound provides continuous, real-time localization control.

The absolute number of pulses and the number of therapy sessions are determined by the therapy concept of the treating physician.

The required energy level can be reached by gradually increasing the energy level depending on the patient's sensitivity to pain and the reaction to analgesic effects caused by shock wave application.

The recommended cumulated positive energy flux density per treatment summed up over all sessions should be approximately 1300 mJ/mm^2.

10.3 Results

Table 10.2. Recently published results of shock wave therapy of plantar fasciitis

Author	Patients	Follow-up (months)	Success rate %	Energy
Hammer 2000	44	6 months	70	3 × 3000, 0.12 mJ/mm^2
Sistermann 1998	34	36 months	67	
Levitt 2000	20	12 months	82	2 × 1000, 16 KV
Buch 1996	110	6 months	80	2000, 0.28 mJ/mm^2
Rompe 1996	50	52 weeks	58	3 × 1000, 0.08 mJ/mm^2
Dahmen 1995	12	12.1 months	83.3	1–10 × 0.08 mJ/mm^2
Odgen 2000	234	12 weeks	76	2 × 1000, 16 KV

For treatment of plantar fasciitis, the low to medium energy treatment regimen is appropriate. Rompe et al. performed the treatment of plantar fasciitis and of heel spur using low energies. They applied 1000 shock waves each in three sessions with an interval of one week at an average energy flux density of 0.08 mJ/mm^2. Of these patients 52% were still free of complaints or had significantly improved after one year. This prospective, randomized study was placebo controlled. Rompe showed that in the placebo-controlled group, which received just 10 shock waves at 0.08 mJ/mm^2, only 10% of the patients achieved comparable results. Dahmen had reported previously that the application of shock waves can lead to an improvement of the clinical status of plantar fasciitis. They treated a small population with a highly variable treatment regimen and reported 83% good to very good results.

Levitt et al. showed in a feasibility study including 20 patients that 1000 shock waves at low energy of 16 kV applied twice can lead to more than 80% good to very good results. In a subsequent multicenter, randomized, placebo-controlled study including 234 patients, a 76% success rate was achieved by Ogden et al. in a subgroup of patients existing essentially of active treatment patients and of subsequently treated placebo patients. The percentages meeting the so-called overall success criterion after 3 months were 47% for active treatment and 30% for placebo (FDA panel meeting, July 2000). In this study the overall success criterion was defined as combination of improvements regarding the investigator assessment of heel pain, the self-assessment of heel pain on walking, the activity score and use of pain medication.

Meanwhile, numerous studies have shown the positive short-term effect of ESWT on plantar fasciitis. The work group of Sistermann published an investigation with a three year follow-up after shock wave therapy resulting in 67% of the patients still showing good to very good results after that time. With regard to the energy flux density applied, there seems to be no clear difference between the low energy and the medium energy form of therapy with respect to the results achieved. Still the results of Buch et al. indicate that with medium energy applications there is a tendency toward achieving slightly improved results. He treated a population of 110 patients once with 2000 shock waves at medium energy with an energy current density of 0.28 mJ/mm^2. Of the patients treated with this regimen, 80% were classified as good to very good.

There were no reports about serious side effects. Occasionally small local petechial skin bleedings and hematomas developed; all of which healed completely.

11 Pseudarthrosis

11.1 Definition

Failure of the bone to unite six months after fracture or after osteosynthetic stabilization of fracture.

11.2 Pathology

Mechanical factors like instability, loss of substance or ingrown soft tissue as well as biological factors (devitalized bone fragments, infections) lead to pseudarthrosis.

11.3 Clinical Diagnosis

Most significant is an inhibited functionality accompanied by pain under pressure and loading with insufficient stability and fixation in the region of the fracture. Painful local edema may also occur.

X-ray: No osseous reconstruction at the site of fracture.

X-ray and scintigraphic diagnosis classifications:
  Hyperthropic pseudarthrosis = biologically active
  Athropic pseudarthrosis = biologically inactive.

11.4 Conservative and Surgical Therapy

Conservative Treatment:
Reset fracture and immobilization with plaster cast or splint, no loading.

Surgery:
Reosteosynthesis, autogenous or homogenous bone graft, internal stabilization (nail), external fixation.

11.5 ESWT

1) Prerequisites

1. Patient should have been treated with conservative therapy for a minimum of six months and have shown no signs of fracture healing.
2. Patient information given and written consent obtained.
3. Exclusion of contraindications.

> **IMPORTANT!**
> Studies conducted to date indicate that
> 1. Good results can be achieved when treating hypertrophic pseudarthrosis with ESWT.
> 2. In the case of atrophic pseudarthrosis after a healed infection, the ESWT success rate is lower.
> 3. Poor results have been reported in the presence of an acute infection at the site of the non-union, in the case of pseudarthrosis with a fracture gap in excess of 5 mm and in case of non-axial alignment of bony fracture endings.
> 4. Surgery remains the treatment of choice for atrophic bone metabolism.

2) Anesthesia

If required general anesthesia (anesthesia mask or intubation), epidural anesthesia or local anesthesia with a local anesthetic agent (Bupivacain, Xylocain).

3) Positioning and Localization

Positioning depends on the method of localization and the anatomical structure to be treated. Patients may be in the prone or supine position. The use of an X-ray system requires the patient to be placed on a treatment table. An integrated X-ray system is preferred to an ultrasound localization system because pseudarthrosis cannot be located accurately with ultrasound in all cases. The patient's limb or extremity should be fixed in a splint during treatment to avoid movement of the patient.

The therapy head can be readjusted using X-ray control during treatment to assure precise coupling.

4) Coupling

Following the application of ultrasound gel to the ROI, the therapy head is precisely focused at the ROI using X-ray control. The therapy head is coupled exactly along the edges of the pseudarthrosis gap after fixation of the affected area (ROI).

5) Treatment Regimen

Shock wave therapy can be initiated following precise X-ray localization.

Treatment should start at energy level 1. The energy level should be increased gradually to the desired treatment level depending on the patient's sensitivity to pain and the reaction to analgesic effects caused by shock wave application.

Correct focusing should be controlled regularly using X-ray imaging.

The number of pulses and the number of therapy sessions applied are determined by the therapy concept of the treating physician. The recommended treatment regimen is 500 to 800 high-energy shock waves per 1 cm gap length resulting in 6,000 to 10,000 shocks per treatment. Results may be seen after 6 weeks to 4 months. No re-treatment is recommended during this period.

Shock wave treatment should be followed by immobilization of the region in a cast, if possible under full load of this extremity, to support the fracture heeling process.

11.6 Results

Table 11.1. Recently published results of shock wave therapy of pseudarthrosis

Author	Patients	Follow up	Success rate (Consolidation) %	Energy
Beutler 1999	27	6 months	41	2 × 2000, 18 kV
Vogel 1997	48	3.4 months	60.4	1 × 3000, 0.6 mJ/mm^2
Gerdesmeyer 2000	35	6 months	62.9	2 × 2000, 0.5 mJ/mm^2
Schleberger 1995	45		89	1 × 3000, 16–18 KV
Diesch 1997	172	3 months	66	1–3 × 2–3000, 0.3 mJ/mm^2
Rompe 1997	58	52 weeks	52	1 × 3000, 0.6 mJ/mm^2
Schaden 1996	40	24 weeks	55	1 × 1–4000, 20–28 KV
Vogel 1997	52	52 weeks	52	1 × 3000, 0.6 mJ/mm^2

Since the first work by Valchanou and Michailov in 1991, the bone inductive effect of shock wave therapy has been known. It has been rapidly adopted for clinical application. This development was justified by scientific results showing a positive effect of shock waves on osteoblasts. In 1995, Schleberger and coworkers showed that the high energy application of shock waves could be used therapeutically for the treatment of pseudarthrosis. In a group of 45 cases of pseudarthrosis, osseous consolidation was obtained in 89% of patients after applying 3000 shock waves at 16–18 KV during one session. Fracture healing reported by Beutler et al. in a study published in 1999 is significantly lower. In a group of 27 cases of pseudarthrosis, bony fusion was achieved in only 41% of the cases. In this case the treatment was also performed at high energies with 2 times 2000 shock waves at 18 KV.

In 1997 Rompe reported a 52% healing rate in a prospective study in which he has treated 58 patients with pseudarthrosis at different localizations with high energies. He applied 3000 shock waves once with an energy flux density of 0.6 mJ/mm^2. Taking account of the results of experimental studies already performed, the use of low energy does not seem to be justified. Gerdesmeyer et al. also applied high energy with an energy flux density of 0.5 mJ/mm^2. In this case ESWT was used in 35 patients with 2 times 2000 shock waves in an interval of 6 weeks achieving an osseous consolidation in 62.9% of the cases. Similarly good results were achieved by the work group of Schaden, Diesch and Vogel with healing rates between 52 and 66% after high energy ESWT.

Numerous authors report side effects in the form of small hematomas and petechial skin bleedings, which occur almost regularly in conjunction with this high energy mode of application. All healed completely.

12 Additional Literature on ESWT of the Musculoskeletal System

Assimos DG, Boyce WH, Furr EG, McCullough DL (1989) Selective elevations of enzyme levels after ESWL. J Urol 142:687–690

Augat P, Claes L, Surger G (1995) In vivo effect of shock waves on the healing of fractured bone. Clinical Biomechanics 10:374–378

Beutler S, Regel G, Pape HC, Machtens S, Weinberg AM, Kremeike I, Jonas U, Tscherne H (1999) Extracorporeal shock wave therapy for delayed union of long bone fractures – preliminary results of a prospective cohort study. Unfallchirurg 102:839–847

Boxberg W, Perlick L, Giebel G (1996) Stoßwellenbehandlung bei therapieresistenten Weichteilschmerzen. Chirurg 67:1174–1178

Brümmer F, Bräuner T, Hülser D (1990) Biological effects of shock waves. World J Urol 8:224–232

Buch M (1998) Clinical results in calcaneal spur. In: Siebert, Buch (eds) Extracorporeal Shock Waves in Orthopaedics. Springer, Berlin Heidelberg New York, p 45

Buch M (1998) Clinical results in tendinosis calcarea. In: Siebert, Buch (eds) Extracorporeal Shock Waves in Orthopaedics. Springer, Berlin Heidelberg New York, p 47

Buch M (1998) Review: Effects and Side Effects of Shock Wave Therapy. In: Siebert, Buch (eds) Extracorporeal Shock Waves in Orthopaedics. Springer, Berlin Heidelberg New York, pp 3–58

Burhenne HJ, Paumgartner G, Ferrucci JT (1990) Biliary Lithotripsy II. Year Book Medical Publishers, Chicago London

Chaussy Ch, Eisenberger F, Jocham D, Wilbert DM (1997) High Energy Shock Waves in Medicine. Thieme, Stuttgart New York

Chaussy Ch, Schmiedt E, Walther V, Brendel W, Forssmann B, Hepp W (1982) Extracorporeal Shock Wave Lithotripsy. In: Chaussy (ed) New Aspects in the Treatment of Kidney Stone Disease. Karger, Basel München Paris

Daecke W, Loew M, Schuhknecht B, Kusnierczak D (1997) Der Einfluss der Applikationsdosis auf die Wirksamkeit der ESWA bei der Tendinosis calcarea der Schulter. Orth Praxis 33:119–123

Dahm K (1996) Shock wave treatment of painful plantar heel spur. User Letter – Dornier-MedTech 1:14–16

Dahmen GP, Franke F, Gonchars V, Poppe K, Lentrodt S, Lichtenberger S, Jost S, Montigel J, Nam VC, Dahmen G (1995) Die Behandlung knochennaher Weichteilschmerzen mit extrakorporaler Stoßwellentherapie. In: Chaussy, Eisenberger, Jocham, Wilbert (eds) Die Stoßwelle – Forschung und Klinik. Attempto, Tübingen, pp 175–186

Dahmen GP, Meiss L, Nam VC, Skruodies B (1992) Extrakorporale Stoßwellentherapie (ESWT) im knochennahen Weichteilbereich der Schulter – Erste Therapieergebnisse. Extracta orthopaedica 15:25–27

Daniel MP, Burns JR (1990) Renal function immediately after piezo-electric ESWL. J Urol 144:10–12

Delius M, Brendel W (1989) Mechanisms of action in extracorporeal shock wave lithotripsy. In: Delius, Brendel, Ferucci (eds) Biliary Lithotripsy. Year Book Medical Publishers, Chicago London, pp 31–42

Delius M, Draenert K, Al Diek, Y, Draenert Y (1995) Biological effects of shock waves: In vivo effect of high energy pulses on rabbit bone. Ultrasound in Med Biol 21:1219–1225

Delius M, Enders G, Xuan Z, Liebich HG, Brendel W (1988) Biological effects of shock waves: kidney damage by shock waves in dogs – dose dependence. Ultrasound Med Biol 14:117–122

Delius M, Ueberle F, Eisenmenger W (1998) Extracorporeal shockwaves act by shockwave-gas bubble interaction. Ultrasound Med Biol 24:1055–1059

Delius M, Ueberle F, Gambihler S (1995) Acoustic energy determines haemoglobin release from erithrocytes by extracorporeal shock waves. Ultrasound in Med Biol 21:707–710

Diesch R, Haupt G (1997) Anwendung der hochenergetischen extrakorporalen Stoßwellentherapie bei Pseudarthrosen. Orthop Prax 33:470–471

Diesch R, Haupt G (1998) Extracorporeal shock wave in treatment of pseudarthosis, tendinosis calcarea of the shoulder, and calcaneal spur. In: Siebert, Buch (eds) Extracorporeal Shock Waves in Orthopaedics. Springer, Berlin Heidelberg New York

Eisenberger F, Miller K, Rassweiler J (1991) Stone Therapy in Urology. Thieme, Stuttgart New York

Eisenberger F, Schmiedt E, Chaussy Ch, Wanner K, Forssmann B, Hepp W, Pielsticker K, Brendel W (1977) Berührungsfreie Harnsteinzertrümmerung. Deutsches Ärzteblatt 74:1145–1150

El Damanhoury H, Schaub T, Stadtbäumer M (1991) Parameters influencing renal damage in ESWL: An experimental study in pigs. J Endo Urol 5:37

Gerdesmeyer L, Bachfischer K, Mittelmeier W, Gradinger R (2000) High energetic extracorporeal shock waves in the therapy of pseudarthrosis; clinical and radiological results. Calcif Tissue Int 66(Suppl 1):125

Gerdesmeyer L, Gradinger R, Rosner S, Bachfischer K (2000) The effect of extracorporeal shock waves on tendinitis calcarea of the shoulder. SIROT Congress Abstracts: 5–6

Graff J, Richter KD, Pastor J (1990) Wirkung von hochenergetischen Stoßwellen auf Knochengewebe. Akt Urol 25:76

Haist J, Steeger D (1994) Die Stoßwellentherapie (ESWT) der Epicondylopathia radialis und ulnaris. Ein neues Behandlungskonzept knochennaher Weichteilschmerzen. Orth. Mitteilungen 3:173

Haist J, von Keitz-Steeger D, Mohr G, Schulze G, Weber F (1998) The orthopaedic shock wave therapy in the treatment of chronic insertion tendopathy and tendinosis calcarea. In: Siebert, Buch (eds) Extracorporeal Shock Waves in Orthopaedics. Springer, Berlin Heidelberg New York, pp 159–163

Haist J, von Keitz-Steeger D (1995) Stoßwellentherapie knochennaher Weichteilschmerzen – ein neues Behandlungskonzept. In: Chaussy et al (eds) Die Stoßwelle: Forschung und Klinik. Attempto, Tübingen, pp 162–165

Haist J (1995) Einsatzmöglichkeiten der analgetisch wirksamen extrakorporalen. Stoßwellentherapie (ESWT) an der Schulter. Orthop Praxis 9:591–593

Haist J (2000) Shockwave treatment for radial and ulnar epicondylitis. Coombs, Schaden, Zhou (eds) Musculoskeletal Shockwave Therapy, Greenwich Medical Media, London, pp 111–113

Hammer DS, Rupp S, Ensslin S, Kohn D, Seil R (2000) Extracorporal shock wave therapy in patients with tennis elbow and painful heel. Arch Orthop Trauma Surg 120:304–307

Haupt G, Katzmeier P (1995) Anwendung der hochenergetischen extrakorporalen Stoßwellentherapie bei Pseudoathrosen, Tendinosis calcarea der Schulter und Ansatztendinosen (Heel spur, Epicondylopathia). In: Chaussy et al. (eds) Die Stoßwelle: Forschung und Klinik. Attempto, Tübingen, pp 143–152

Jakobeit C, Welp L, Winiarski B, Schuhmacher R, Osenberg T, Splittgerber T, Spelsberg G, Buntrock W, Missulis U, Kroll U, Schmeiser A, Beer M, Watzlawik A, Olschner G, Winarski B (1998) Ultrasound-guided extracorporeal shock wave therapy of tendinosis calcarea of the shoulder, of symptomatic plantar calcaneal spur (heel spur) and of epicondylopathia radialis et ulnaris. In: Siebert, Buch (eds) Extracorporeal Shock Waves in Orthopaedics. Springer, Berlin Heidelberg New York, pp 166–180

Johannes EJ, Kaulesar Sukul DMKS, Bijma AM, Mulder PGH (1994) Effect of high energy shock waves on human fibroplasts in suspension. Journal of Surgical Research 57:677–681

Karlsen SJ , Berg KJ (1991) Acute changes in renal function following ESWL in patients with a solitary functioning kidney. J Urol 145:253–256

Krischek O, Pompe JD, Hopf C, Vogel J, Herbsthofer B, Nafe B, Burger RZ (1998) Extracorporeal shockwave therapy in epicondylitis humeri ulnaris or radialis – a prospective, controlled, comparative study. Orthop Ihre Grenzgeb 136:3–7

Levitt R, Alvarez R, Odgen J (2000) FDA studies of musculoskeletal shockwave therapy for lateral epicondylitis and heel pain syndrom. Eds. Coombs, Schaden, Zhou. Musculoskeletal Shockwave Therapy, Greenwich Medical Media, London, pp 107–110

Liedl B, Jocham D, Lunz C, Schuster C, Chaussy C (1989) Prävalenz und Inzidenz der arteriellen Hypertonie bei ESWL-behandelten Nierensteinpatienten. Urologe A 28:130–133

Lingeman JE, McAteer JE et al (1988) Bioeffects of extracorporeal shock wave lithotripsy. Urol Clin North Am 15:507–514

Lingeman JE (1992) Bioeffects of ESWL: to worry or not? J Urol 148:1025

Loew M, Daecke W, Kusnierczak D, Rahmanzadeh M, Ewerbeck V (1999) Shock-wave therapy is effective for chronic calcifying tendinitis of the shoulder. J Bone Joint Surg Br 81:863–867

Loew M, Jurgowski W, Mau H, Perlick L, Kusnierczak D (1995) Die Wirkung extrakorporal erzeugter hochenergetischer Stoßwellen auf den klinischen, röntgenologischen und histologischen Verlauf der Tendinosis calcarea der Schulter – eine prospektive Studie. In: Chaussy et al (eds) Die Stoßwelle: Forschung und Klinik. Attempto, Tübingen, pp 153–156

Loew M, Jurgowski W, Mau HC, Thomsen M (1995) Treatment of calcifying tendinitis of rotator cuff by extracorporeal shock waves: a preliminary report. J. Shoulder Elbow Surg 4:101–106

Maier M, Lienemann A, Refior H (1997) Gibt es magnetoresonanztomographische Veränderungen nach Stoßwellenbehandlung bei Tendinitis calcarea? Z Orthop 135:20–21

Maier M, Ueberle F, Rupprecht G (1998) Physikalische Parameter extrakorporaler Stoßwellen. Biomedizinische Technik 43:269–274

Newman DM, Coury T, Lingeman JE (1986) ESWL experience in children. J Urol 136:238–240

Odgen J, Alvarez R, Levitt R, Cross GL (2000) Chronic heel pain: Results of FDA shockwave study. 3rd Congress of the ISMST Naples, Abstracts, p 51

Roessler W, Steinbach P, Nicolai H, Hofstaedter F, Wieland W (1993) Effects of high-energy shock waves on the viable human kidney. Urol Res 21:273–277

Rompe JD, Burger R, Hopf C, Eysel PJ (1998) Shoulder function after extracorporal shock wave therapy for calcific tendinitis. Shoulder Elbow Surg, Sep; 7:505–509

Rompe JD, Eysel P, Hopf C, Krischek O, Vogel J, Bürger R, Jage J, Heine J (1997) Extrakorporale Stoßwellentherapie in der Orthopädie. Fort d Med 115:26–33

Rompe JD, Hopf C, Küllmer K, Witzsch U, Nafe B (1996) Analgesic effect of extracorporeal shock wave therapy on chronis tennis elbow. J. Bone Joint Surg 78(B):233–237

Rompe JD, Hopf C, Küllmer K, Witzsch U, Nafe B (1996) Extrakorporale Stoßwellentherapie der Epicondylopathia humeri radialis – ein alternatives Behandlungskonzept. Z Orthop 134:63–66

Rompe JD, Hopf C, Nafe B, Bürger R (1996) Low-energy extracorporeal shock wave therapy for painful heel: a prospective controlled single-blind study. Arch Orthop Trauma Surg 115:75–79

Rompe JD, Hopf C, Rummler F (1994) 2 Jahre Extrakorporelle Stoßwellentherapie (ESWT) in der Orthopädie – Indikationen und Resultate. Orth Mitteilungen 3:173

Rompe JD, Küllmer K, Riehle HM, Herbsthofer B, Eckardt A, Bürger R, Nafe B, Eyse P (1996) Effectiveness of low energetic extracorporeal shock waves for chronic plantar fasciitis. J Foot Ankle Surg, pp 215–221

Rompe JD, Küllmer K, Vogel J, Eckardt A, Wahlmann U, Eysel P, Hopf C, Kirkpatrick CJ, Bürger R, Nafe B (1997) Extrakorporale Stoßwellentherapie – Experimentelle Grundlagen, klinischer Einsatz. Orthopäde 26:215–228

Rompe JD (1997) Extrakorporale Stoßwellentherapie, Grundlagen, Indikationen Anwendungen. Chapman & Hall, Weinheim, p 36

Rompe JD (1996) Stoßwellentherapie: Therapeutische Wirkung bei spekulativem Mechanismus. Z Orthop 134:13–19

Schaden W, Meznik A, Russe F, Pachucki A (1996) Anwendung der extrakorporalen Stoßwellentherapie (ESWT) bei 40 Patienten mit Pseudarthrosen. Sonderdruck der Firma HMT: 1–22a

Schleberger R, Dahm K, Werner T (1997) Two-center comparison of extracorporeal shock wave therapy (ESWT) in calcaneal spur. In: Chaussy et al. (eds) High Energy Shock Waves in Medicine. Thieme, Stuttgart New York, pp 117–120

Schleberger R, Diesch R, Schaden W, Rompe J D, Vogel J (1997) Four-center result analysis of extracorporeal shock wave treatment of long bone non-unions. In: Chaussy et al (eds) High Energy Shock Waves in Medicine. Thieme, Stuttgart New York, pp 112–116

Schleberger R, Senge T (1992) Non-invasive treatment of long-bone-pseudarthrosis by shock waves (ESWL). Arch Orthop Trauma Surg 111:224–227

Schleberger R (1995) Anwendung der extrakorporalen Stoßwelle am Stütz- und Bewegungsapparat im mittelenergetischen Bereich. In: Chaussy, Eisenberger, Jocham, Wilbert (eds) Die Stoßwelle – Forschung und Klinik, pp 166–174

Schräbler S (1999) Ein abtastendes Verfahren zur Darstellung und Analyse von Stoßwellen in Flüssigkeiten. Shaker, Aachen

Singh VR, Adya VP, Aftab A, Sanjay Y (1990) A stress save propagation technique for bone repair study. IEEE Transactions on Biomedical Engineering 37:1014–1017

Sistermann R, Katthagen BD (1998) 5-years lithotripsy of plantar heel spur: experiences and results – a follow-up study after 36.9 months. Z Orthop Ihre Grenzgeb 136:402–406

Smits G, Jap P et al. (1994) Biological effects of high energy shock waves in mouse skeletal muscle: correlation between ^{3}P magnetic resonance spectroscopy and microscopic alteration. Ultrasound Med Biol 19:399

Steinbach P, Hofstaedter F, Nicolai H, Roessler W, Wieland W (1993) Determination of the energy-dependent extent of vascular damage caused by high-energy shock waves in an umbilical cord model. Urol Res 21:279–282

Ueberle F (1997) Acoustic parameters of pressure pulse sources used in lithotripsy and pain. In: Chaussy et al (eds) High Energy Shock Waves in Medicine. Thieme, Stuttgart New York, pp 76–85

Ueberle F (1995) Internationale Normungsaktivitäten für Lithotriptermessungen. In: Chaussy et al (eds) Die Stoßwelle. Attempto, Tübingen, pp 52–58

Ueberle F Pressure Pulses in Extracorporeal Shock Wave Lithotripsy (ESWL) and Extracorporeal Shock Wave Pain Therapy. ESWT contribution to a review on shock wave applications in medicine. Springer, Berlin Heidelberg New York (to be published)

Ueberle F (1997) Shock Wave Technology In: Siebert, Buch (eds) Extracorporeal Shock Waves in Orthopaedics. Springer, Berlin, pp 59–87

Ueberle F (1999) Stoßquellenparameter für die ESWT und ihre Interpretation, Abstracts of the 4th Kasseler Stoßwellen-Symposium, March

Ueberle F (1999) Stoßwellentechnik. Giebel (ed). Barth, Heidelberg, pp 3–37

Valchanou VD, Michailov P (1991) High energy shock waves in the treatment of delayed and nonunion of fractures. Int Orthop (SICOT) 15:181–184

Van Arsdalen KN, Kurzweil S, Smith J, Levin R (1991) Effect of lithotripsy on immature rabbit bone and kidney development. J Urology 146:213–216

Vergunst H, Terpsta OT, Brakel K, Lameris JS, Nijs HGT, ten Kate FWJ, Schröder FH (1989) Safety and efficacy in biliary lithortripsy. In: Burhenne HJ, Paumgartner G, Ferrucci JT (eds) Biliary Lithotripsy II. Year Book Medical Publisher, Chicago London, pp 25–28

Vogel J, Hopf C, Eysel P, Rompe JD (1997) Application of extracorporeal shock waves in the treatment of pseudarthrosis of the lower extremity. Preliminary results. Arch Orthop Trauma Surg 116:480–483

Vogel J, Rompe JD, Hopf C, Heine J, Bürger R (1997) Die hochenergetische extrakorporale Stoßwellentherapie (ESWT) in der Behandlung von Pseudarthrosen. Z Orthop 135:145–149

Wess O, Ueberle F, Dührßen RN, Hilcken D, Krauß W, Reuner Th, Schultheiß R, Staudenraus J, Rattner M, Haaks W, Granz B (1997) Working Group Technical Developments – Consensus Report. In: Chaussy et al. (eds) High Energy Shock Waves in Medicine. Thieme, Stuttgart, pp 59–71

Wess O, Ueberle F, Staudenraus J, Rattner M, Haaks W, Granz B (1995) Entwicklung eines Gewebephantoms für vergleichende Messungen an Lithotriptoren. In: Chaussy et al (eds) Die Stoßwelle. Attempto, Tübingen, pp 30–36

Wilbert DM, Strohmaier WL, Flüchter, SH, Bichler KH (1989) Urinary proteins as parameters of renal function changes after ESWL. Investigative Urology 3. Eds. Rübben H et al Springer, Berlin Heidelberg New York, pp 249–253

Wolf T, Breitenfelder J (1998) Course observations after extracorporeal shock wave therapy (ESWT) in cases of pain in the locomotor system with circumscribed localization. In: Siebert, Buch (eds) Extracorporeal Shock Waves in Orthopaedics. Springer, Berlin Heidelberg New York, pp 181–188

Yang C, Heston WDW, Gulati S, Fair WR (1988) The effect of high energy shock waves (HESW) on human bone marrow. Urol Res 16:427–429

Yeaman LD, Jerome CP, McCullough DL (1989) Effects of shock waves on the structure and growth of the immature rat epiphysis. J Urol 141:670–674

Yokoyama M, Shoji F, Yanagizawa R (1992) Blood pressure changes following ESWL for Urolithiasis. J Urol 147:553

13 Additional Literature, Video Tapes and CD-ROMs on Ultrasound Examination of the Musculoskeletal System

Arbeitskreis Stütz- und Bewegungsorgane der DEGUM (1996) (Deutsche Gesellschaft für Ultraschall in der Medizin) Protocol of the meeting of the work group Musculoskeletal System of DEGUM, 01. 20.

Bauer G, Burri C, Swobodnik W, Rübenacker S (1987) Meniskussonographie. Dtsch Z Sportmed 38:74–80

Bauer G, Heuchemer T, Haas S (1989) Sonographisches Bild der Meniskusläsionen. Ultraschall 10:198–201

Bauer G, Rübenacker S (1988) Sonographische Meniskusdarstellung: Welcher Schallkopf ist geeignet? Ultraschall 9:48–51

Baumann D, Kremer H (1977) Arthrographie und Sonographie in der Diagnostik von Bakercyst. Fortschr Röntgenstr 127:463–466

Behrend R, Hinzmann J, Heise U (1988) Sonographische Darstellung von Kreuzbändern und deren Läsionen. Orthop Praxis 7:459–462

Billy de M, Doucet J, Quentin G (1975) Angular dependence of the backscattered intensity of acoustic waves from rough surfaces. Conference Proc. Ultrasonics International

Casser H R, Sohn C, Kiekenbeck A (1990) Current evaluation of sonography of the meniscus. Results of a comparative study of sonographic and arthroscopic findings. Arch Orthop Trauma Surg 109:150–154

Casser HR, Füsting M (1991) Meniskus. In: Harland U, Sattler H: Ultraschallfibel Orthopädie, Traumatologie, Rheumatologie. Springer, Berlin Heidelberg New York

Casser HR, Prescher A, Füsting M, Tenbrock F (1990) Analyse möglicher Fehlinterpretationen in der Meniskussonographie anhand sonoanatomischer Untersuchungen. Orthop Praxis 12:813–818

Correll J (1988) Die operative Verlängerung/Korrektur durch Fixateur externe bei angeborener oder erworbener Gliedmaßendeformität. Med Orth Technik 108:135–143

Correll J (1988) Operative Korrektur verkürzter oder deformierter Gliedmaßen mit der Ilzarov-Methode. Kinderarzt 19:1261–1277

Dragonat P, Claussen C (1980) Sonographische Meniskusdarstellungen. Fortschr Röntgenstr 133:185–187

Füsting M, Casser HR (1991) Dynamische Untersuchungstechnik in der Meniskussonographie. Sportverl Sportschad 5:27–36

Graf R, Schuler P (1995) Sonographie am Stütz- und Bewegungsapparat bei Erwachsenen und Kindern. 2. Auflage, Chapman & Hall, London Glasgow Weinheim

Grifka J, Richter J (1992) Meniskussonographie. Enke, Stuttgart

Gruber G, Konermann W, Diepolder M, Harland U (1992) Sonographie der Stütz- und Bewegungsorgane. Teil 1: Sonoanatomie (Lehrfilm). Springer, Berlin Heidelberg New York

Gruber G, Konermann W, Diepolder M, Harland U (1995) Sonographie der Stütz- und Bewegungsorgane. Teil 2: Typische pathologische Befunde (Lehrfilm). Springer, Berlin Heidelberg New York

Gruber G, Konermann W, Harland U (1994) Sonographische Standardschnittebenen am Kniegelenk. Ultraschall in Klinik und Praxis 9:47–51

Gruber G, Konermann W, Müller-Miny H, Gruber GM (1997) Standardisierte sonographische Untersuchung des Kniegelenkes. Ultraschall in Med 18:52–61

Gruber G, Konermann W (1997) Sonographie der Stütz- und Bewegungsorgane. Standardschnittebenen nach den Richtlinien der DEGUM. Chapman & Hall, London Glasgow Weinheim

Gruber G, Martens D, Konermann W (1998) Stellenwert der sonographischen Untersuchung bei Läsion des Ligamentum collaterale mediale. Z Orthop, S 337–342

Gruber G, Nebe M, Bachmann G, Litzlbauer HD (1998) Die sonographische Untersuchung in der Diagnostik der fibularen Bandruptur. Eine vergleichende Studie: Sonographie versus Röntgenuntersuchung. Fortschr Röntgenstr 169:152–156

Gruber G, Stürz H (1996) Postoperative sonographische Kontrolle der Distraktionsosteotomie. OP-Journal, S 321–325

Gruber G (1996) Sonographie des Kniestreckapparates. Ultraschall in Med 17:206–211

Harland U, Diepolder M, Gruber G, Knöss H-P (1991) Die sonographische Bestimmung des Humerusretrotorsionswinkels. Z Orthop 129:36–41

Harland U, Dittrich A, Gruber G (1992) Die Bestimmung der Tibiatorsion. Ultraschall in Klinik und Praxis 7:54–59

Harland U, Gruber G (1995) Muskeln. In: Graf, Schuler (eds) Sonographie am Stütz- und Bewegungsapparat bei Erwachsenen und Kindern. 2. Edition. Chapman & Hall, London Glasgow Weinheim, pp 319–334

Harland U, Gruber G (1995) Sehnen. In: Graf, Schuler (eds) Sonographie am Stütz- und Bewegungsapparat bei Erwachsenen und Kindern. 2. Edition. Chapman & Hall, London Glasgow Weinheim, pp 335–353

Harland U, Sattler H (1991) Ultraschallfibel Orthopädie, Traumatologie, Rheumatologie. Springer, Berlin Heidelberg New York

Harland U (1988) Die Abhängigkeit der Echogenität vom Anschallwinkel an Muskulatur und Sehnengewebe. Z Orthop 126:117–124

Harland U (1986) Die sonographische Untersuchung des Schultergelenkes. Med Orthop Techn 106:48–52

Harland U (1987) Schultersonographie. Ultraschall Klin Prax 2:10–18

Harland U (1987) Sonographische Bestimmung des Retrotorsionswinkels am Humerus. Orthop Prax 23:626–631

Heckmatt JZ, Dubowitz V, Leemann S (1980) Detection of pathological change in dystrophic muscle with B scan ultrasound imaging. Lancet 1:1389–1390

Hedtmann A, Fett H, Moraldo M (1987) Ultraschalldiagnostik der Schulter bei Sportverletzungen. Dtsch Z Sportmed 38:86–98

Hedtmann A, Fett H (1991) Atlas und Lehrbuch der Schultersonographie. Bücherei des Orthopäden, 2.Ed. Enke, Stuttgart, p 52

Hedtmann A, Weber A, Schleberger R, Fett H (1986) Ultraschalluntersuchung des Schultergelenkes. Orthop Praxis 9:647–661

Hien N M, Schricker T, Wirth CJ (1987) Sonographische Funktionsdiagnostik bei Kapselbandverletzungen des Kniegelenkes. Hefte zur Unfallheilkunde, pp 1083–1085

Hien NM, Sedlmeier P, Schricker T (1986) Sonographische Diagnostik bei Kapselbandverletzungen des Knie- und Sprunggelenkes. In: Otto R, Schnaars P (eds) Thieme, Stuttgart New York, Ultraschalldiagnostik 85:558–559

Hien NM, Röhr E, Sohn C (1988) Kniegelenk. In: Graf, Schuler (eds) Sonographie am Stütz- und Bewegungsapparat bei Erwachsenen und Kindern. VCH Verlagsgesellschaft, Weinheim, pp 217–261

Hinzmann J, Behrend R, Heise U (1988) Sonographische Beurteilung typischer Läsionen bei der Schulterluxation. Z Orthop 126:570–573

Huppertz R, Pfeil J, Kaps HP (1990) Sonographische Verlaufskontrollen von Verlängerungsosteotomien. Orthop 128:90

Kaarmann H (1995) Ultraschalltechnik. In: Graf R, Schuler P (eds) Sonographie am Stütz- und Bewegungsapparat bei Erwachsenen und Kindern. 2. Edition. Chapman & Hall, London Glasgow Weinheim, pp 1–37

Kaftori JK, Rosenberger A, Pollack S, Fish JH (1977) Rectus sheath hematoma: ultrasonographic diagnosis. AJR 128:283–285

Kainberger F, Nehrer S, Breitenseher M, Seidl G, Baldt M, Rand T, Imhof H (1996) Die Sonomorphologie der Achillessehne und ihre Differentialdiagnose. Ultraschall in Med 17:212–217

Keßler T, Duchêne W, Köpke J, Winkler H, Wentzensen A (1997) Ist die Arthrosonographie des Ellenbogens eine sinnvolle Ergänzung bei der Beurteilung des traumatisierten Gelenkes? Akt Traumatol 27:158–161

Konermann W, Gruber G, Freund MC, Scheller G, Harland U (1994) Standardisierte sonographische Untersuchung des Ellenbogengelenks. Erläuterung der sonographischen Standardschnittebenen durch korrespondierende MRT-Schnitte. Ultraschall in Klinik und Praxis 9:163–165

Konermann W, Gruber G, Gaa J (2000) Standardisierte sonographische Untersuchung des Hüftgelenks. Ultraschall in Med 21:137–141

Konermann W, Gruber G, Gruber GM, Freund MC, Ellermann A, Harland U (1996) Standardisierte sonographische Untersuchung des Schultergelenks. Erläuterung der sonographischen Standardschnittebenen durch korrespondierende MRT-Schnitte. Ultraschall in Klinik und Praxis 10:124–129

Konermann W, Gruber G (1993) Die dorsalen Anteile der lumbalen Wirbelsäule im Sonogramm – interventionelle Möglichkeiten. Ultraschall in Klinik und Praxis 8:175

Konermann W, Gruber G (1992) Interventionelle Sonographie am Stütz- und Bewegungsorgan. Ultraschall in Klinik und Praxis 7:150

Konermann W, Gruber G (1997) Septische Coxitis im Kindesalter. Sonographische Differentialdiagnosen. Orthopäde 26:830–837

Konermann W, Gruber G (1998) Sonographische Standardschnittebenen an der oberen Extremität – Schulter – und Ellenbogengelenk. Ultraschall in Med 19:130–138

Konermann W, Gruber G (2000) Ultraschalldiagnostik der Stütz- und Bewegungsorgane. Thieme, Stuttgart

Konermann W, Gruber G (1993) Ultraschallgeführte Punktion, Injektion und Biopsie am Stütz- und Bewegungsorgan. Ultraschall in Klinik und Praxis 8:175

Konermann W, Mailänder W, Gruber G, Hettfleisch J, Bettin D, Klein D, Güth V (1995) Die sonographische Beinlängen- und Beinlängendifferenzmessung. In: Graf R, Schuler P (eds) Sonographie am Stütz- und Bewegungsapparat bei Erwachsenen und Kindern. 2. Edition. Chapman & Hall, London Glasgow Weinheim. Z Orthop 133:442–452

Konermann W, Pellegrin de M (1993) Die Differentialdiagnose des kindlichen Hüftschmerzes im Sonogramm. Orthopäde 22:280–287

Kramps HA (1979) Einsatzmöglichkeiten der Ultraschalldiagnostik am Bewegungsapparat. Z Orthop 117:355–364

Krappel F, Schmitz R, Harland U (1997) Die sonographische Diagnostik der vorderen Syndesmosenruptur. Z Orthop 135:116–119

Kresse H (1969) Der Einfluss des Einfallswinkels bei der Ultraschallechodiagnostik. Elektromedizin. Special Edition, pp 43–45

Küllmer K, Eysel P, Rompe J (1995) Erste Erfahrungen der intraoperativen Sonographie bei lumbalen Bandscheibenvorfällen. Ultraschall in der Medizin; special edition 1:46

Kupatz P, Hinzmann J (1995) LWS-Sonographie in der Rehabilitation. Ultraschall in der Medizin; special edition 1:46

Malzer U, Feltes E, Schuler P, Griss P (1989) Ultraschallartefakte in der Meniskussonographie. Ultraschall Klin Prax 4:171–176

Malzer U, Kienapfel H, Schuler P (1988) Möglichkeiten und Grenzen der sonographischen Darstellung des Meniskus und angrenzender Strukturen im Kniegelenk. Ultraschall Klin Prax 3:141–145

Marcelis S, Daenen B, Ferrara MA (1996) Normal and pathologic regional ultrasound: lower extremities. Knee: Cruciate ligament injury. In: Dondelinger (ed) Peripheral musculoskeletal ultrasound atlas. Thieme, Stuttgart New York

Matter HP, Gruber G, Konermann W, Gruber GM, Litzlbauer HD (1998) Standardisierte sonographische Untersuchung des Sprunggelenkes. Erläuterung der sonographischen Standard-Schnittebenen durch korrespondierende MRT-Schnitte. Ultraschall in Med 19:34–39

Matter HP, Gruber G, Konermann W, Harland U (1995) Möglichkeiten der sonographischen Diagnostik bei Acromioclavikulargelenk-Verletzungen Typ Tossy III im Vergleich mit gehaltenen Röntgenaufnahmen. Sportverletzung – Sportschaden 9:14–20

McDonald DG, Leopold GR (1972) Ultrasound B-scanning in the differentiation of Baker's cyst and thrombophlebitis. Br J Radiol 45:729

Melzer C (1995) Artefakte bei der Schultersonographie. In: Graf R, Schuler P (eds) Sonographie am Stütz- und Bewegungsapparat bei Erwachsenen und Kindern. 2. Edition. Chapman & Hall, London Glasgow Weinheim, pp 81–93

Nebe M (1996) Stellenwert der sonographischen Untersuchung in der Diagnostik fibularer Bandrupturen. Med Diss, Gießen

Röhr E (1987) Die Sonographie des Kniegelenkes. Orthopädische Praxis 11:939–943

Röhr E (1987) Sonographische Darstellung des vorderen Kreuzbandes. Ultraschall in der Medizin 8:37

Röhr E (1985) Experimentelle Untersuchungen zur sonographischen Darstellung der Kreuzbänder. Fortschr Röntgenstr 143:4

Röhr E (1988) Kniegelenkssonographie. Thieme, Stuttgart New York

Röhr E (1985) Sonographische Darstellung des hinteren Kreuzbandes. Röntgenblätter 38:377

Röhr E (1989) Sonographische Untersuchungen zur Innenmeniskus-Hinterhornläsion. Orthop Praxis 11:728–733

Sattler H (1995) Ellbogen und Hand. In: Graf R, Schuler P (eds) Sonographie am Stütz- und Bewegungsapparat bei Erwachsenen und Kindern. 2. Edition. Chapman & Hall, London Glasgow Weinheim, pp 125–140

Schmid A, Schmid F, Tiling T (1988) Einsatz der Arthrosonographie bei der Diagnostik von Tossy-Verletzungen am Schultergelenk. Aktuell Traumatol 18:134–138

Schmid A, Schmid F, Tiling T (1988) Stellenwert der Arthrosonographie beim Lachman-Test. In: Hofer, Glinz, Beck (eds) Arthroskopie bei Instabilität des Kniegelenkes. Enke, S 27–37

Schricker T, Hien NM, Wirth CJ (1987) Klinische Ergebnisse sonographischer Funktionsuntersuchungen bei Kapselbandläsionen am Knie- und Sprunggelenk. Ultraschall in der Medizin 8:27

Schricker T (1986) Sonographische Funktionsuntersuchungen bei Kapselbandläsionen am Knie- und oberen Sprunggelenk. Med Diss, München

Schwarz W, Hagelstein J, Minholz R, Schierlinger M, Danz B, Gerngroß H (1997) Manuelle Sonometrie des Kniegelenks. Unfallchirurg 100:280–285

Selby B, Richardson ML, Montaria MA, Teitz CC, Larson R, Mack LA (1987) High resolution sonography of the menisci of the knee. Invest Radiol 21:332–335

Sell S, Balensiefen F, Küsswetter W (1997) Sonographie von Achillessehnenläsionen – eine experimentelle Studie. Ultraschall in Med 18:124–129

Sell S, Esenwein S, Gaismaier C, Moosmaier J, Küsswetter W (1997) Die Sonographie der Patellasehne – eine experimentelle Studie. Z Orthop 135:261–265

Sell S (1997) Ultraschall Stütz- und Bewegungsorgane – Interaktive CD ROM. Enke, Stuttgart

Sohn C, Bastard G (1992) Dreidimensionale Ultraschalldarstellung. Dtsch med Wochenschr 117:467–472

Sohn C, Casser HR (1988) Meniskussonographie. Springer, Berlin Heidelberg New York

Sohn C, Gerngroß G, Griesbeck F (1987) Wertigkeit, Technik und klinische Anwendung der Meniskussonographie. Unfallchirurg 90:173–179

Sohn C, Gerngroß G, Meyer P, Sohn G (1987) Meniskussonographie. Aussagekraft und Treffsicherheit im Vergleich zur Arthrographie und Arthroskopie oder Operation. Fortschritte Med 105:81–85

Subject Index